# POWER TRIP

## ALSO BY MICHAEL E. WEBBER

*Thirst for Power*

# POWER TRIP

## THE STORY OF ENERGY

———

## MICHAEL E. WEBBER

BASIC BOOKS
New York

Basic Books
Hachette Book Group
1290 Avenue of the Americas, New York, NY 10104
www.basicbooks.com

Printed in the United States of America

First Edition: May 2019

Published by Basic Books, an imprint of Perseus Books, LLC, a subsidiary of Hachette Book Group, Inc. The Basic Books name and logo is a trademark of the Hachette Book Group.

The Hachette Speakers Bureau provides a wide range of authors for speaking events. To find out more, go to www.hachettespeakersbureau.com or call (866) 376-6591.

The publisher is not responsible for websites (or their content) that are not owned by the publisher.

Print book interior design by Trish Wilkinson.

Library of Congress Cataloging-in-Publication Data

Names: Webber, Michael E., 1971– author.
Title: Power trip: the story of energy / Michael E. Webber.
Description: First edition. | New York, NY: Basic Books/Hachette Book Group, 2019. | Includes bibliographical references and index.
Identifiers: LCCN 2018047339| ISBN 9781541644397 (hardcover) | ISBN 9781541644380 (ebook)
Subjects: LCSH: Power resources. | Energy development. | Energy consumption. | Technology and civilization.
Classification: LCC TJ163.2 .W368 2019 | DDC 333.79—dc23
LC record available at https://lccn.loc.gov/2018047339

ISBNs: 978-1-5416-4439-7 (hardcover); 978-1-5416-4438-0 (ebook)

LSC-C

10 9 8 7 6 5 4 3 2 1

*This book is dedicated to the more than one billion people who do not have access to modern forms of energy, but need it, and those who are trying to make our energy system better.*

# CONTENTS

Prologue

# THE STORY OF ENERGY

Energy is amazing. It brings light to dark and warmth to cold. It preserves our food, cleans our water, makes our cities safe, and lets us roam the world. To put it more starkly, energy is an invisible builder of civilization. Maybe *the* builder of civilization. Invisible, but no less real. Many archaeologists use materials—stone, bronze, iron—to denote the eras of human evolution in world prehistory, but changes in energy might be a better boundary for more recent human history.[1] Nearly every major social, political, scientific, and cultural revolution in the last few hundred years is at least partially a result of advances in our ability to harness energy.

Energy managed wisely—with long-term sustainability, affordability, and security in mind—gives us health and wealth; managed unwisely, it makes us sick and poor. Despite its importance, energy is out of sight and usually out of mind. We can't see energy, we don't often contemplate it unless we're thinking specifically of oil or electricity, we don't even necessarily know what it is, yet if one day it were no longer available to us in the form, price, and abundance we expect, its absence could cause a global social collapse.

And our relationship with energy is perpetually changing. We are in the midst of an energy transition yet again as some countries seek to reduce greenhouse gas emissions and others seek to improve energy access. Those policies, combined with global population growth, economic growth, urbanization, motorization, industrialization, and electrification, mean we are changing how much energy we use, how we use it, and what forms we use.

The stakes are high during this transition because energy is unique: no other physical factor in society has such a wide-ranging impact on public health, ecosystems, the global economy, and personal liberties. Energy is a key enabler of entrepreneurship and innovation, the economic engines for society. Energy holds much promise for humanity, but a lack of energy can doom populations to incredible suffering. Because energy is such an important driver of modern living, safe and reliable access to energy has become a precondition of mobility and affluence. If we run out of energy, food will spoil, water will become unavailable or contaminated, and it will be hard to get to work if we still have jobs. Good decisions will require an honest look at its challenges as the billion-plus people who lack access to electricity, sanitation, or safe water seek a pathway out of poverty. And that requires first understanding how energy weaves its way through our daily lives.

Because problems like climate change span continents, for the first time in human history the world must come together to face collective energy challenges. That we can come together in the first place is because energy enables us to do so. Energy is the medium in which globalism exists, so it would be only natural that energy requires global solutions. On the table are decisions about how to invest trillions of dollars to build and reinvent an energy system that ensures access to clean, affordable, reliable energy. These decisions, which must be made in a matter of years to stave off the worst effects of climate change, will have far-reaching consequences for billions of people over many decades to come.

To make wise decisions about the future of energy, we must first understand its importance to everything we care about. Energy is hidden yet ubiquitous, embedded in the world around us. While much of the popular debate hinges around which forms of energy we should use or prohibit, it is the end-uses that define our modern society. In the end, we care less about the exact form of our energy and instead want clean water, preserved food, a chance for a high quality of life (financial freedom, so we can have comfortable homes and things to do), mobility so we can move about our lives, and the peace of mind that comes with a sense of safety. Those end-uses and benefits—water, food, transportation, wealth, cities, and security—are what we seek, and they are all tied to energy.

## THE MAGIC OF ENERGY

By classical definition, energy is the capacity to do work. That is, energy is the ability to do interesting and useful things. It is the potential to harvest a crop, refrigerate it, or fly around the world. The corollary is thus that a lack of energy is the inability to do work. Access to energy enables our civilization. An absence of energy inhibits it. Energy is also conserved: it is inherently a finite resource. We cannot make more of it; we can only move it around or transform it. At its heart, our relationship with energy is about harnessing the benefits and containing the environmental impacts of those transformations.

Despite the importance of energy, strictly speaking we still don't know what it is. We know it's real, and we can measure its presence. But we can't see it or touch it. Or even explain it. Nobel laureate Richard Feynman admitted, "It is important to realize that in physics today, we have no knowledge of what energy is."[2] Energy is mysterious, powerful, and invisible at the same time.

Famed science fiction writer Arthur C. Clarke once issued his Third Law, noting that "Any sufficiently advanced technology is indistinguishable from magic."[3] This law emerged from a series of essays about the future he wrote out of exasperation because of his sense that the scientific community was widely pessimistic about our collective ability to innovate. Clarke was an optimist and believed in the almost magical capability of advanced technologies. That magic is our modern energy system, and Clarke cited many examples of remarkable energy-related advances to prove his point.

Energy's magic quietly brings illumination, information, heat, clean water, abundant food, motion, comfort, and much more to our homes and factories with the flick of a switch or the touch of a button. Leonardo da Vinci said that "by its power [the energy of fire] shall transform almost everything from its natural condition into another," acknowledging the transformative power of energy. The goal of this book is to demystify the magic of energy and to celebrate its transformative power.

In the United States each year, 1 quadrillion British thermal units (BTUs) of energy ingested in the form of food is needed for the basal metabolism of our 325 million bodies. Yet we collectively consume 100 quadrillion BTUs of energy. That means we are consuming far more energy than we need just for sustenance. That additional energy lets us heat, cool, move, and do things for the equivalent of ourselves and ninety-nine other people. It's as if we are all emperors now commanding ninety-nine slaves (in the form of kilowatts and BTUs) to do our bidding for us. Energy gives us the ability to live in modern cities with skyscrapers, which require elevators, elevated water, and indoor lighting—all powered by electricity—to operate. Today farmers in the United States produce more than twice as much per acre as American farmers did at the end of World War II.[4] Energy—in the form of irrigation from electric pumps, fertilizers from natural gas, pesticides from petroleum, and mechanization in the form of diesel-powered tractors and harvesters—enabled this transformation. Today we have an abundance of food in countries where energy access is universal.

We use energy to give us mobility, both social mobility from getting richer by consuming energy for our factories or other industrial purposes, and physical mobility when moving from place to place powered by gasoline and jet fuel. To our ancient ancestors who spent weeks traveling from one city to another, our modern pace of transportation—spanning a continent in a few hours and the globe in less than a day—would appear as magical as teleportation in a sci-fi movie seems to us. Energy is at the heart of this magical pace.

## THE STORY OF ENERGY IS THE STORY OF CIVILIZATION

For thousands of years the story of energy was slow-moving and incremental, but in the last few hundred years in the developed world, the energy story has become more interesting. It's the developed world that demonstrated the benefits of increasing energy access and it's the developed world that is grappling with how to maintain those benefits while reducing the environmental consequences. Older examples of energy consumption are useful for context, but the crux of the decision at hand—how to expand energy access without scorching the earth—is more recent. Up until a few hundred years ago, water was the most important ingredient for civilization, but ever since the industrial revolution, which gave us the combination of advanced fuels and new machines, energy has joined water as one of the two most important ingredients of civilization. Prior to that, the history of energy is thinner—there's less to say about those centuries because our relationship with energy was fairly stable. That makes the industrial revolution a worthy boundary for a discussion of energy.

Just as we use carbon radioisotopes to date ancient organic materials and artifacts, we can use energy as a dating element for the modernity of various societies. The type of energy used and the technologies available are markers of time. Older, less developed societies used primitive solid fuels such as cow dung, wood, charcoal, or straw. Burning these fuels provided the energy needed to heat homes, cook food, and create household items by making lime, boiling soap, and so forth. The heat could also be used to forge and soften metals. Those primitive fuels also became products and building materials. For example, wood—a form of bioenergy, which is just energy from the sun stored in biomatter from years of photosynthesis—gives us firewood for heating; lumber for buildings, ships, and fences; and pulp for paper.

Older civilizations also used flowing water, wind, and animals to perform work and move people and goods. The classic European image of a windmill or medieval water wheel that powered a mill to grind grain, saw wood, or polish glass is a cliché indicating an older society.

As time went on, the fuels became higher-performing. Whale oil was a popular illuminant that displaced dirtier, dimmer candles and torches. Despite that, whale oil was eventually displaced by kerosene for indoor lighting, which was cheaper and brighter and didn't smell pungent like burning blubber. Coal, which has even higher energy density, burns hotter, and produces less smoke than wood or cow dung, became a preferred fuel for indoor heating and industry. That means oil saved the whales and coal saved the forests. Coal's higher combustion temperatures were a natural fit for metalworking, which meant that the rise of coal enabled the rise of steel that enabled the development of better tools. In a virtuous cycle, the importance of coal kick-started a mining boom in England that led to deeper and deeper mines. As those mines filled with water, the need for a pump became acute. The steam engine—powered by burning coal—was invented by Thomas Newcomen and later improved by James Watt, who were motivated by the need to pump the water out of the mines. Solving the water problem enabled even greater coal production.

That same steam engine also allowed, for the first time in history, the conversion of heat into motion. Heat is easy to produce by the simple act of burning fuels. But motion always depended on water power (such as that from flowing streams), wind power (for sailboats or windmills), or muscle power (from walking under our own strength or from animals). But with the steam engine, burning a fuel could give the energy needed to boil water to create steam that would drive pistons to create motion. This breakthrough eventually led to the creation of a self-reinforcing loop of energy and innovation, as ironclad steam trains powered by coal from their hoppers and riding on steel rails forged in smelters that used coal could transport coal dug with modern tools from faraway mines.

The steam engine, miraculous as it was for its day, would eventually be supplanted by the internal combustion engine. While bulky steam engines powered by coal were a good fit for heavy vehicles like mining equipment, ships, and trains, they were awkwardly sized for vehicles at the scale of an individual or a family. Gasoline-powered engines—much lighter and more powerful than their steam-powered cousins—were a good fit for transportation, enabling a revolution of personal mobility. The impact of individualized, motorized transportation is hard to fathom. In total, it created some of the world's largest industries: automotive manufacturing (using steel made from coal and tires made from petroleum at factories operating on electricity), oil production for automotive fuels, and road construction (using asphalt from petroleum and cement made with coal). These breakthroughs changed where and how we lived, as transportation encouraged far-flung yet interconnected societies and electricity enabled urban living that reached to the sky.

Whether we realize it or not, the forms of energy and their related technologies mark the passage of time. If moviemakers or artists want to visually depict antiquity or medieval times, then they will show someone lifting water from a well by hand, riding a horse, or burning wood. To show the industrial revolution of the late 1800s, they will show factories, trains, and ships burning coal with black smoke curling from the chimneys. To show the twentieth century, gasoline-powered cars and electric appliances paint the image.

Energy innovation is also used as a timestamp that defines our eras. Energy conversion systems like the steam engine created the industrial revolution, the turbine engine enabled the jet age, and the electrical transistor enabled the information age. That these human eras are marked by energy advances makes perfect sense because energy drives individual metabolisms in living creatures and the urban metabolism of modern civilizations.

Some might argue that scientific discovery or modern medicine is the true marker of progress. But energy is even more fundamental. Medical professionals would have trouble saving lives without scalpels and other instruments made from metal forged with fossil fuels, lighting to allow surgeons to see clearly, medicines made from petrochemicals, plastic devices made from natural gas, and electricity to heat water for disinfection.

Energy is more than just fancy tools and cozy comforts of life: democratization of energy access reduces economic inequity. Countries where citizens have access to energy are freer. Looking at the earth at night, speckled with lights from cities, shows where people have access to electricity. The darkness of North Korea despite its proximity to brightly lit South Korea reveals that autocracy and energy poverty are different expressions of the same idea. Energy is freedom because it means people can move from one place to another and because lighting is available so students can study at night. Getting an education is the best pathway to economic opportunity. It also enables citizens to get information that improves the functioning of a democracy or resistance to autocracy.

Critically, energy can be used to improve water supplies. Once energy and water were available in abundance, food could be grown on a large scale, achieving higher yields per acre of farmland despite a drop in the number of agricultural laborers. As food productivity went up, workers could specialize in other pursuits. That specialization, combined with the availability of energy, also made it easier for many people to accumulate wealth. Energy was used for running factories, including the machines themselves and the devices such as lighting and air conditioning that made indoor settings comfortable and productive. In addition, the advent of electric lighting reduced the cost and improved the brightness of home illumination, which made it easier for students to do their lessons, just as students could stay in school longer because they weren't as necessary on the farm.

Because agriculture was now able to feed an entire nation with a small fraction of the population—thanks to energy—the allure of rural living was offset by the limited economic opportunities there as agriculture advanced its technological capabilities.

Consequently, many people—especially youth—moved to cities to look for jobs and to use their newfound wealth and freedoms for the entertainment and cultural activities available in urban areas. The additional spare time that energy bought could be spent on cultural or frivolous delights such as listening to the opera or watching movies, improving quality of life.

Living in cities required modern transportation to bring food, goods, and people back and forth. Personal mobility was the perfect companion to social mobility, bringing the freedom of movement to a rising number of people. It allowed city workers to live in the suburbs or to take plane or boat trips for work or vacation. These transportation systems include coal-fired trains, gasoline-powered cars, and spaceships propelled by rocket fuels.

Energy is also intermingled with personal, urban, and national security. Through public lighting, energy brought safety to dark cities by illuminating the shadows where serial killers and other predators hid. It enabled weapons of war—nuclear bombs and the planes that dropped them. Energy is also a cause of war, a target of war, and a strategic driver of a war's key decisions. The intermingling of energy security and national security, a driving part of national defense strategy in the first part of the twenty-first century after the terrorist attacks of September 11, 2001, highlighted the connections between Western oil consumption and threats to our well-being. As energy became the central enabling feature of modern civilizations, it also became its greatest security vulnerability.

Bringing these pieces together, it is clear that the story of energy is the story of civilization.

## ENERGY AS A GLOBAL GRAND CHALLENGE

We must acknowledge from the outset that if energy is the builder of civilizations, it can also be their destroyer. Richard

Smalley, a brilliant, Nobel prize–winning chemist, spent the last few years of his life educating the world on what he considered to be the grandest obstacles for the world. Smalley prioritized the top ten problems for humanity as follows: (1) energy, (2) water, (3) food, (4) environment, (5) poverty, (6) terrorism and war, (7) disease, (8) education, (9) democracy, and (10) population.[5] Energy is at the top of the list because its availability is the key to unlocking the next nine challenges. Only if we solve the problem of ensuring access to clean, reliable, affordable energy can we solve our water problems. With abundant water and energy we will have the irrigation, fertilizers, and tractors we need to grow, harvest, and distribute food so that food security problems can be solved. Cleaning up our energy is critical to avoiding and mitigating environmental impacts. But it is hard to prioritize environmental protection when people are hungry, which is why food is a higher priority than the environment. Making energy accessible will elevate people out of poverty, and reducing inequality reduces the risks of terrorism and war. Energy can be used to refrigerate food and disinfect water, reducing the risk of disease. In many parts of the world, women and girls spend 1.4 hours every day collecting fuelwood, which keeps them out of school.[6] Using modern energy to make those chores less laborious gives girls a chance to go to school. Electric lighting lets them study at night. Education is a primary ingredient of successful democracies, and educated women tend to have fewer children, so increasing energy access improves education, enhances democracy, and slows population growth.

Using Smalley's logic, energy is the single most important opportunity in the world and also its most important problem. In addition, there is vast energy inequity. The average worldwide citizen consumes half the energy of a British resident, who consumes half the energy of an American, who consumes two-thirds the energy of a typical Texan. If global population keeps growing and the world's residents want the magic of energy at the same level of consumption as a Texan—with bustling factories, comfortably air-conditioned houses, and big cars—then global energy production and consumption will need to expand by an order of magnitude. Given the level of national security and environmental impacts of today's global energy system, it is hard to imagine expanding its production, movement, and use by a factor of ten without a major transformation.

Bringing the benefits of energy at this scale to everyone without scorching the atmosphere, acidifying the oceans, or denuding land will require a new kind of magic. That is our global challenge.

# Chapter 1

# WATER

We begin the story of energy with water, because it is with water that life and civilization begin. Ensuring access to water is the first priority for individuals and societies because water is so critical to life. We need it to nourish our bodies and to grow our crops. At the cellular level, we need water for our body's circulation system. Without water, our bodies would shut down and we would die of dehydration. If that did not kill us, then we would ultimately die of starvation, anyway, because the foods we eat also need water to grow. We also would not have the fibers we need—no cotton, wool, or leather—to clothe and shelter ourselves from the weather.

In many places around the world, getting water—for ourselves and our fields—takes energy to lift, move, and treat it. Since the arrival of advanced fuels and machines in the industrial revolution, energy joined water as one of the two most important ingredients for civilization. In fact, lifting and treating water was one of society's first priorities for energy use. Because the modern water system directly depends on energy, that means modern civilization depends on energy and water. It goes the other way, too: the energy systems depends on water up and down its supply chain.[1] This interconnection has good news and bad news associated with it.

For millennia, before advanced forms of energy were available, water was the most important basis for civilization.[2] It has even been suggested that water is the motivating reason to organize humans into larger societal groups; that is, there is no reason to form a society except for the need to collectively manage water resources.[3] The Chinese have noted this connection with their word *zhi*, which means "to rule" and "to regulate water." An article in the *Economist* noted that "the Chinese word for politics (zhengzhi) includes a character that looks like three drops of water next to a platform or dyke. Politics and water control, the Chinese character implies, are intimately linked."[4]

In Classical Nahuatl, which is the language of pre-Columbian Aztec society, the word for city, *altepetl*, means "water mountain," by joining *atl* (water) and *tepe-tl* (mountain). So the Aztecs considered control of water a key ingredient and enabling step to forming a city.

Beautiful aqueducts and public fountains were among the defining elements of the Roman empire. Roman water infrastructure was an obvious symbol of their dominance. As they conquered a new territory, they would Romanize it by building waterworks to project their power.[5]

If the good news is that water abundance and collaboration can foment civilization, then the bad news is that scarcity and conflict can cause societies to fail. Researchers studying caves in China concluded that multi-decade droughts occurred at the end of three of the five long-lasting Chinese dynasties—the Tang (618–907), Yuan (1271–1368), and Ming (1368–1644)—implying that water strain was a trigger for dynastic collapse. Drought has also been associated with the collapses of the Khmer empire in the thirteenth century, the Mayans in 900 CE, and the Ancestral Puebloans of the American desert Southwest (whom the Navajo and others knew as the Anasazi).

Other examples abound. The MENA region (Middle East and North Africa) seems to suffer perpetual civil unrest. Water resources are strained there, which crimps food supplies, exacerbating the situation. The global Syrian refugee crisis of the mid-2010s can be tied directly to widespread bankruptcy of rural farmers whose crops were destroyed by drought.[6] The displacement of Syrian farmers caused by that epic drought led to a flood of refugees crossing into Europe, triggering a global political backlash partially credited for leading to the election of Donald Trump and the famed Brexit vote in 2016. In that way, it could be said that drought toppled the prevailing governments in the United States and England, half a world away.

Water is also critical to public health. Over a billion people still do not have access to improved water or wastewater systems. As Molly Walton, an analyst at the International Energy Agency in Paris, noted in her commentary on World Water Day in 2018, "Energy has a role to play in achieving universal access to clean water and sanitation."[7] If we want to solve our public health problems, we need to solve our water problems. And if we want to solve our water problems, then we need to solve our energy problems. Importantly, doing so is critical to achieve the UN's sustainable development goals.

The opportunity presented by energy's and water's interdependence is that infinite availability of one means we can have infinite availability of the other. With unlimited energy, our water problems will be solved because we can drill deeper water wells, build longer aqueducts, or desalinate seawater to quench our thirst. With unlimited water, all our energy problems will be solved because we can dam up rivers to make hydroelectric power or grow biofuels in the desert. But those solutions have their own environmental impacts to worry about, such as silting up the rivers, causing massive runoff into watersheds, or finding a way to dispose of a lot of brine left over from desalination. Regardless, we do not live in a world of infinite energy and water; we live in a world of constraints. Because of their close relationship, a constraint in one becomes a constraint in the other. And a shortage can send effects rippling across civilization.

## WATER FOR MECHANICAL POWER

Before we used water for modern energy systems, water was used for mechanical power and for transportation. Anywhere water flowed reliably with a significant drop, the force of the falling water could be harnessed. The potential energy of the water at higher elevation could be converted into mechanical energy: As it fell, it would rotate a large wooden wheel that could in turn power a series of shafts, wooden or metal gears, and axles connected to a tool that would perform some useful task. Waterwheels on flowing water gave the power to polish glass, grind grain, spin spindles, operate bellows for metalworking, and saw wood. This mechanical power supplemented and amplified what was available from the muscle power of human laborers or domesticated animals.

Water was also used for transportation. Rivers, lakes, and oceans had been used for transportation for millennia, and ultimately canals—or manmade waterways—were created to facilitate the movement of people and goods. Water for transportation and water for manufacturing coupled nicely. Water power was harnessed through waterwheels to manufacture goods, and then canals let those products move to customers easily and efficiently.

Those different implements converged to turn America's waterways into powerhouses to drive a modernized economy, sometimes in the same location. In Lowell, Massachusetts, a savvy proprietor managed the flow of water using a system of dams, locks, and canals and then sold off a slice of the potential energy of the falling water to manufacturers who wanted it to drive their mills and factories.[8]

Because the United States has abundant hydropower potential, its industrialization was powered by water. According to water expert Martin Doyle, "settlers of the eighteenth and nineteenth century built their villages around small dams powering waterwheels," and "the power of the Susquehanna River was as essential to grinding colonial grain as the Merrimack River was to spinning the fabric of New England textile mills."[9] This was in contrast with England, which had abundant and easily accessible coal but poor hydropower potential. Consequently, its manufacturing was mobilized by steam power rather than water power. "Colonists all along the East Coast initially put waterwheels to work in mills to process timber, which was essential for building settlements and one of the key raw materials that was plentiful in America but in short supply in Europe. By the time the earliest sawmills were built in England in the 1660s, several hundred were already being used in colonial New England." Ultimately, "sawmills and gristmills were the centerpiece of the colonial economy."[10]

Those gristmills were significant amplifiers of productivity for grinding wheat into flour and corn into meal. Humans required about two days of labor to grind a bushel of wheat into flour; horses could do the same work in a few hours. But a typical eighteenth-century water-powered gristmill could grind dozens of bushels of flour or cornmeal daily.[11]

Unsurprisingly, processed materials are more valuable than raw ones. Lumber is worth more than timber, and flour is worth more than wheat. The modern analogies are that gasoline is worth more than crude oil and chemicals are worth more than natural gas. Industrious humans used energy to upgrade their

natural resources into higher-value commodities they could export elsewhere. And, because of the higher value density of the
finished products, it was smart to do so. Flour was more valuable
per pound and easier to transport than wheat. The same could
be said for the water that goes into it. It made more sense to
transport a pound of flour than the 1,000 pounds of water required to grow it. Water and energy made it possible to process
a wide range of goods, creating value along the way.

While hydropower was a major advance over muscle power,
it still had its drawbacks. Namely, you needed incredible quantities of water flowing down altitude drops to make it work.
And in many parts of the world—including England, as already
noted, and large swaths of the United States—nature did not
provide that combination. And this is where steam power became revolutionary.

Burning fuels to make heat to boil water allowed heavy machinery to be moved by the force of steam. The industrial age
is really the age of steam, as the invention of steam engines
created the opportunity to turn heat into motion. Creating heat
is rudimentary and the materials for it are readily available, but
the ability to turn heat into motion was revolutionary.

In the United States, this transition occurred at the end of
the 1800s. The amount of water needed for steam was a fraction
of the water needed for hydropower. And, even in low-lying
areas that didn't have the altitude drop needed for waterwheels,
there was sufficient water to make steam. Steam not only increased the energetic output over hydropower but also freed
manufacturers to build their factories where they wished.

Wood could generate the heat needed for boilers to make
steam, but coal was a much better fuel, generating more heat
with less pollution and at lower cost. In this way, fuels—and in
particular, fossil fuels—freed us from the rivers, allowing us to
move to more convenient locations. Hydropower cared if the

land was flat, but steam power did not care. Consequently, major cities such as Chicago, Cleveland, and Detroit emerged in flat areas on the edge of lakes. Their topography would not accommodate hydropower, but they had access to easy shipping from the lakes and were close to many raw materials. Just as flowing water shaped the geographic story of industrialization, energy changed the landscape of industrialization so that it could take place almost anywhere.

## WATER FOR ENERGY

Moving from direct water use to steam was liberating, but another major advance awaited: electrical power. Using water to generate electricity would give even more flexibility for manufacturers, as it is possible to move electricity many hundreds of miles over transmission lines, whereas moving water or steam is very cumbersome.

The electrical age was just as transformative as the steam age. Though Benjamin Franklin's famous experiment in the 1700s was an eye-catching illustration of the similarities between lightning and the sparks of static electricity, most of the electricity experiments in the eighteenth century and the early nineteenth century had been lab-scale benchtop exploration to satisfy scientific curiosity. And they tended to use low-voltage, direct current devices, such as small batteries or fuel cells. It was not until the late 1800s that larger-scale alternating current systems enabled useful appliances such as motors and the affordable incandescent light bulb and made electricity a more valuable part of day-to-day life. Like the rise of the information economy in the late 1900s, the rise of electrification was very rapid once it passed the tipping point.

Electricity can be generated many ways, but among the simplest is by spinning magnets around a coil to induce a current.

And, as had already been known for centuries, water could easily rotate a wheel. Flowing water turned overshot waterwheels, which rotated a shaft that could be used to power equipment such as rotating blades at sawmills, rolling stones at gristmills, and spindles at textile factories. The same concept could be applied to spinning magnets to make electricity.

The world's first hydroelectric power plant was built in 1879 along the Fox River to light hundreds of bulbs in Appleton, Wisconsin. The first hydroelectric dams were not very large and were more reminiscent of the medieval overshot waterwheels used for mechanical power.

These smaller hydroelectric power plants started to proliferate in the late 1800s to meet the demand for light bulbs and small motors for industrial work. Some, like the plant built at Niagara Falls in 1882, simply diverted some of the natural flowing force of the water above the falls. But at other locations, where water was not already falling hundreds of feet over a ledge, a dam was needed to create a reservoir at a higher elevation than the water body below it. Early dam builders might have created a small reservoir that elevated water ten feet above the water body below, but generally speaking, these structures were not considered to have that great an impact on the river's natural flow. Over time, larger dams were built, and for multiple reasons: for power, flood control, navigation, and irrigation.

By the end of the 1800s, factories had begun to electrify. In 1900, only 4 percent of Chicago's factories were electrified; thirty years later, it was 78 percent.[12] These dams and their affordable, powerful electricity gave the United States a competitive economic advantage.

In Ireland in the 1920s, the Electricity Supply Board hatched a scheme to create a national grid fueled by hydroelectric power from the River Shannon as a way to catch up with America's electric factories.[13] Around the same time, Oskar von Miller in Germany was building a hydroelectric power system outside Munich that used the natural 200-meter height difference of two lakes to build the Walchensee Hydroelectric Power Station, which is still operational today.

Although the allure of reliable and cheap electricity certainly helped promote the spread of dams, their rise in popularity can also be attributed to other reasons. As the United States expanded and its population increased, economic activity increased in floodplains and the losses from floods also grew. Because those floods were really destructive, they helped kick off a dam- and levee-building boom to manage flood risks. As dams were built, powerhouses could be included, providing electricity as a handy by-product of infrastructure built to control the flow of water.

Perhaps there was no more compelling case for dam building than the 1927 Mississippi River flood, which drove home the dangers of a river jumping its banks. The losses were staggering: as much as $1 billion at a time when the federal budget was typically less than $3 billion; over 700,000 people lost their homes and as many as 300,000 were rescued from houses, rooftops, levee crowns, and even trees. Water expert Martin Doyle observed that as a consequence of these killer floods, "river systems became highly-engineered, optimized hydraulic machines. Early twentieth-century floods gave the motivation, the progressives gave the ideology, and the New Deal provided the resources."[14] That is, several different forces converged to turn America's waterways into powerhouses to drive a modernized economy.

Not much later, the military buildup for World War II created a massive demand for the electricity generated from dams that were built to mitigate flooding. With the country still in the shadow of the Great Depression, jobs were scarce, and large water projects were a way to keep people working while achieving the other useful benefits. In response, the dam build-out in the United States accelerated in the 1930s and continued for a few decades. During this period, the Hoover Dam (initially named the Boulder Dam) and several other prominent dams such as the Shasta and the Grand Coulee were built.

Most of the early build-out in the United States was in the Pacific Northwest's Columbia River basin and in the southeastern United States. The abundant electricity provided by these dams kicked off a large military effort located right next to the massive power plants: aluminum production. Because of World War II, which relied on airplanes more than any prior war in history, there was significant new demand for aluminum. Since aluminum is produced electrolytically from bauxite (by contrast, steel is produced thermally from iron ore), many aluminum smelters were located near dams. Abundant electricity enabled aluminum production at a pace that had never been seen before.

In addition, there was significant demand for enriched uranium for nuclear weapons. Since uranium is enriched with electrically driven centrifuges, the appetite for power was enormous. At one point during the peak of the war effort, 1 percent or more of national electricity consumption was dedicated just to enriching uranium.[15] Dams were a key piece of that effort, and consequently the main nuclear labs for uranium processing were established in Washington State near the Columbia River dams and in Tennessee near the dams built by the Tennessee Valley Authority (TVA). It is telling that some of the Department of Energy's main nuclear processing labs in the United States are still located in those same places: Pacific Northwest National Lab in Washington and Oak Ridge National Lab in Tennessee.

The TVA is a unique quasi-governmental agency that simultaneously manages water and power to serve a region, similar in a way to what the Lower Colorado River Authority does in central Texas or what the Bonneville Power Administration does in the Pacific Northwest. The TVA was formed in 1933 partly as an outgrowth of the women's suffrage movement. After women earned the right to vote in 1920, female activists formed the League of Women Voters to address other important issues. One of their achievements was their instrumental role in form-ing the TVA.[16] They supported it because they wanted jobs, but they also wanted electricity to be available to households—that is, women—in the Deep South.[17] Along with the right to vote, electricity and electric appliances were another expression of freedom.

The TVA came to life from a World War I dam that the War Department had sought to build in Muscle Shoals, Alabama. Today it is a large agency that serves a sprawling territory with a mixture of many dams, coal, and nuclear power.

Muscle Shoals lies along the Tennessee River, between the musical cities of Memphis and Nashville. Music buffs will recog-nize Muscle Shoals as the site of the FAME Studio, where some of the world's most notable acts—Aretha Franklin, Etta James, Percy Sledge, the Rolling Stones, and the Allman Brothers—went to record some of their most famous hits.[18] The connec-tion between music and dams extends beyond Muscle Shoals to America's greatest folksinger. Because the reservoirs created by dams would flood entire valleys, scar the local ecosystem, inhibit fish migration, and in some cases displace a lot of people, there was resistance to dam construction. To help overcome it, the Bonneville Power Administration launched a public relations campaign touting the benefits of hydropower. They made films and printed posters and even commissioned the folk musician Woody Guthrie, who wrote "This Land Is Your Land," to pen a collection of songs about the dams along the Columbia River.[19] Songs like "Roll On, Columbia" and "Song of the Grand Cou-lee Dam" were effective: the dams got built.

Modern hydroelectric dams remain appealing because they are clean at the point of generation, efficient, robust, simple, and start up quickly. By comparison, thermal power plants that burn coal or use heat from nuclear reactions to boil water take many hours or even days to reach full capacity. One of the unfortunate secrets of the modern grid is that for most power plants—almost all nuclear, coal, and natural gas power plants—the power has to already be on before a power plant can be turned on. This creates a remarkable chicken-and-egg situation. What happens if the power goes out because of a storm or equipment failure? A small fraction of power plants are "black start rated," which means they can turn on even if the power is off. Dams are black start rated because even if there is a blackout, gravity always works. In fact, after blackouts, dams are often used to provide the power that lets other power plants turn on and connect to the grid. In that way, water backstops the entire modern economy.

Though dams are relatively clean at the point of generation and therefore popular among stakeholders who wish to limit emissions of greenhouse gases, they are not free of environmental impingement. Their construction has significant ecosystem impacts. Flooding large valleys to create the reservoirs can displace people and irrevocably change the geography. Dams also disrupt fish migration, which has cascading impacts on those who depend on the fish for life and livelihoods. Because of their performance benefits, dams are still desirable, but because of their drawbacks, the construction of major dams is nearly impossible in the United States and Europe, where there is well-organized opposition to them.

But because water projects are a hallmark of a civilized society, they are popular among ruling classes as symbols of political power built by and named for politicians. At the famous palace of Versailles, outside Paris, a large-scale water tower and hydraulic system were built to provide water to tens of thousands of residents while also powering the fountains, which were intended to be fabulous displays of wealth and to demonstrate that water systems could be beautiful as well as functional.

In the nineteenth century, Abraham Lincoln ran on a platform of enhanced water infrastructure—namely, canals—for navigation and commerce. Doyle observed that after the 1927 Mississippi River flood, "Flood control infrastructure projects became a favorite flavor of political pork for the next half century."[20] The largest dam in the world at the time it was built was eventually named for President Herbert Hoover. Even local regional dams—the Buchanan dam outside Austin, Texas, named for a local congressman, for example—fall prey to the same vanity. This convergence was immortalized by Wendell Wilkie, the top executive at the forerunner of today's Southern Company, a massive utility in the South, who was the Republican nominee for president in 1940. He lost to Franklin Delano Roosevelt, who won a third term in the presidency, and whose legacy includes a swath of energy and water infrastructure systems.

The political incentive to build water infrastructure is not unique to the United States. In Asia, Africa, and South America, dams remain popular as a way to electrify a region while securing political power because they promise multifaceted benefits. For example, the Sardar Sarovar Dam in India was designed to provide irrigation for a million farmers, drinking water for 30 million people, 1.5 gigawatts of power, and jobs for five thousand employees.[21]

Hydroelectric power plants can be absolutely massive, both in area and in power generation. The largest power plant in the world, the Three Gorges Dam in China, has a capacity of 22 gigawatts, about the size of twenty nuclear power plants. The gargantuan Hoover Dam, whose proximity to Las Vegas and scenic backdrop makes it a typical tourist destination, is only

2 gigawatts by comparison. Those early dams from the late 1800s have an electrical generating capacity about a thousandth the size of the Three Gorges dam.

The Three Gorges Dam was pursued by Chinese leaders for decades, and it was finally constructed in the 2000s. It is hard to fully appreciate the scale of this dam: It is so large, the reservoir it created is as long as Great Britain.[22] The mass of the water in the reservoir is so significant, it slowed the earth's rotation. By elevating nearly 40 billion tons of water to hundreds of meters above sea level, the dam has essentially made the earth a little fatter in the middle and flatter at the top, extending the day by six-hundredths of a microsecond.[23] If you are ever late for a meeting again, you can blame the Three Gorges Dam for messing up your clocks.

The scale and risk of the Three Gorges Dam, in terms of both its water and its power, are enormous. It is the ultimate testament to human hubris. While it has helped to reduce the risks of flood-related disasters and improved the navigability of the Yangtze River, its creation flooded entire valleys and towns. Geologists worry about the earthquakes and underwater landslides that the water causes as the soft, soaked soils around the reservoir settle to accommodate the new load. In the first decade of its operation, the dam triggered more than five hundred earthquakes with a magnitude greater than 2.0 on the Richter scale, and more than four hundred landslides.

If the Three Gorges Dam were to collapse, it would put approximately 15 million downstream lives or more at risk.[24] In the event of a collapse, a wave would move quickly down the canyons, making it difficult for people to escape. Unfortunately, more than 600 dams are either built, under construction, or in planning in the seismically active Himalayas, putting those dams at serious risk of failure. If the Tehri Dam in India collapses, scientists expect it would produce a wall of water 200 meters high that would put 2 million people at risk.[25] While we can hope such a catastrophe will not happen, unfortunately, dams collapse every once in a while. The near-miss with the Oroville Dam in California in February 2017 was a reminder that one big rainfall can be a triggering event that strains reservoirs to the breaking point. Because of snowmelt and a lot of rain, the Oroville Dam's reservoir overtopped the spillway, which eroded badly. The dam did not break, but the risk of it doing so forced the evacuation of nearly 200,000 people. When dams do break, the results can be horrific. A series of deadly dam failures in the 1970s was part of the inspiration for President Jimmy Carter to create the Federal Emergency Management Agency (FEMA) in 1979.[26]

Though the Three Gorges Dam is already a world-changing power plant, a bigger one is already under consideration: the Grand Inga Dam has been proposed on the Congo River in the Democratic Republic of the Congo. The Congo River is the second largest in the world by volume, with an average flow rate of 1.5 million cubic feet every second falling 300 feet through the falls where the dam has been proposed. The scale of the project is breathtaking: a proposed 40 gigawatts of capacity, nearly twice the size of the Three Gorges Dam. However, it would mostly be a run-of-river project, meaning a simple diversion could be used and a reservoir is not necessary, though one design includes a small reservoir to increase the height of the water's fall. It also holds out the prospect of increasing the African continent's installed generating capacity by 40 percent. Increasing electricity access so significantly in Africa might have nontrivial benefits in terms of enabling industrial activities that would create jobs and wealth while lifting people out of poverty. Though those benefits are attractive, the project is controversial for all the same reasons plaguing other dams: flooding of farmland, ecosystem impacts, and displacement of tens of thousands of people.[27] Because of the rich ecological setting, creating the reservoir would lead to methane emissions from anaerobic decomposition of plant matter below water and might increase the population of mosquitoes or other vectors. It would also trap sediments, which would affect the fertility of downstream fields. The high price tag—nearly $100 billion—in a country rife with corruption is eye-catching, raising the specter of politically connected industries benefiting from the power while poorer communities feel the negative impacts of the dams without accessing electricity. Today the project is at a standstill, though a small portion of it has already been developed.

Water is critical to other parts of the system, too. In addition to mechanical power from waterwheels or electrical power from hydroelectric turbines, water boiled into steam from the heat of atomic reactions or burning coal, gas, oil, or wood spins turbines that generate nearly three-fourths of all the electricity in the world. Then water from lakes, rivers, and oceans is used to cool those same steam power plants to improve their efficiency. Typical homes in the United States use about 20–40 kilowatt-hours of electricity each day, which means 300–600 gallons of cooling water are required daily to make electricity for those homes. That same home would typically use 150 gallons per day for washing, cooking, and drinking (not including watering lawns) for a few residents.[28] That means we use more water at home for our lights and outlets than our faucets and showerheads.

The mining process for resources such as coal, uranium, and rare earth metals used in power systems relies on water to leach out the desired materials. In fact, it was the need for water during the western mining boom of the 1800s that established water as a property right in the western United States.[29] In the eastern United States, people did not own water; rather, riparian law established that they had a right to use a reasonable amount of water adjacent to their land. But when mining activities created demand for water far away from the rivers, water as a material right that could be owned got decoupled from the physical property where it originated.

We also use water to produce fuels. Irrigation is critical to the growth of energy crops such as corn. Water to make steam is also used for the biorefining and fermentation processes that turn corn into ethanol for our cars. Oil and gas are extracted with techniques such as waterflooding or hydraulic fracturing. In waterflooding, water is injected into reservoirs to push the oil and gas out. In hydraulic fracturing, water laced with chemicals and proppants (like sand or miniature ceramic beads) is injected at high pressure into formations to fracture the shale and hold the cracks open so oil and gas can flow out of them. Then fuels are refined or upgraded using steam as a carrier of heat. Lastly, water is used for transporting fuels. Coal moves along barges through inland waterways, and the global trade in oil, refined products, and liquefied natural gas moves by ships across the seas.

In total, about 10 percent of the water consumed in the United States goes to the energy sector.[30] At local scale, water use for oil and gas production can be a much larger fraction, which can pit municipalities against industry in a competition for the most valuable use of water. In some places, the shale revolution puts pressure on watersheds because water is required for hydraulic fracturing in one location in the watershed, and then downstream more water is needed to make steam at a petrochemical facility that turns the natural gas from shale into higher-value chemical products.

Because energy depends critically on water up and down its supply chain, if the water is not available when, where, and how it is needed, then the energy sector might suffer a catastrophic failure. There are many examples around the world where droughts, floods, heat waves, or freezes disrupt the energy system, creating power outages and fuel shortages with life-or-death consequences.

Some unexpected good news, however, is that fossil fuel combustion actually creates water. Hydrocarbons are composed of different chains of hydrogen (H) and carbon (C) atoms attached to each other. For example, the chemical formula for methane is $CH_4$, propane is $C_3H_8$, and gasoline can be approximated as $C_8H_{15}$. When these hydrocarbon fuels react with the oxygen ($O_2$) in air, two major combustion products result: carbon dioxide ($CO_2$) and water vapor ($H_2O$). Because carbon dioxide traps infrared radiation, hydrocarbon combustion unavoidably produces greenhouse gases unless scrubbers are included to avoid emissions. But that water vapor is also created and added to the global hydrologic cycle. Our research at the University of Texas at Austin calculated that global fossil fuel combustion puts more than 12 billion tonnes of water into the atmosphere each year.[31] That is a small fraction of the total mass of atmospheric water vapor, and not enough to affect climate change, but it is nevertheless a nontrivial amount and it means fossil fuel usage actually creates water, which is counterintuitive.

The energy activity enabled by water has been transformative but also has its downside: namely, pollution. It is a great irony that energy lets us treat and clean water, but that energy production also puts water quality at serious risk. Energy's pollution of water takes many forms. One of those is thermal pollution. For example, since steam-fired power plants use water for cooling, in the process, they heat that water up before returning it to rivers, lakes, and oceans. Doing so puts aquatic life at risk if the water is returned at too hot a temperature. Energy systems can also cool the water to unsafe temperatures. Some types of dams release water downstream that is much cooler, disturbing the reproductive cycles of fish. And some gasification terminals receive liquefied natural gas (LNG), which is stored and moved in ships at a very low temperature of −260 degrees Fahrenheit (−160 degrees Celsius). In comparison, the surrounding ocean is much warmer and can be used as a source of heat to gasify the LNG for onshore pipelines. In the process, the adjacent ocean water cools significantly as its heat is transferred to the LNG.

Using water to produce energy also risks serious chemical pollution. The same Tennessee Valley Authority that got its start building dams had a very large-scale coal ash spill around Christmastime 2008. This spill flooded a valley with 5.4 million cubic yards of coal ash, blanketing 300 acres and two dozen homes, creating extensive damage, and requiring an expensive cleanup process. Nearly a decade later, a headline noted "180 New Cases of Dead or Dying Coal Ash Spill Workers" from the rosters of men who labored on the cleanup and in so doing were exposed to dangerous chemicals.[32] Over thirty workers died in the cleanup process in the first decade after the spill, and more deaths are expected.

Other environmental impacts on water include radiation contamination from leaking materials at nuclear facilities and naturally occurring radioactive materials (NORM) that are sometimes produced alongside oil and gas. There are also the extensive mining tails from extractive industries, oil spills, run-off of fertilizers from biofuels production, and the chemical injection associated with hydraulic fracturing. All told, energy can be a dirty business.

For a long time, it was common practice for companies to simply dump their wastes untreated into streams and rivers. I find it very embarrassing as an engineer that in the early 1900s the engineering community supported this program: an editorial in the *Engineering Record* in 1903 declared that it was "more equitable" for upstream cities to discharge waste into a stream and for those downstream to treat it "than for the former city to be forced to put in sewage-treatment works."[33] As a consequence, almost all of the collected wastewater in the United States in the early 1900s was released into streams and lakes as raw, untreated sewage. This belief that industry should be able to dump its waste unscrubbed into rivers is similar to what industry says today about dumping their waste (in the form of $CO_2$ emissions) unscrubbed into the atmosphere.

That pollution can be stunning. The Cuyahoga River was at one time so polluted that it would regularly catch fire. Generally speaking, it is unnerving when water burns. The Cuyahoga River ran through the heart of industrial activity in the Ohio River valley, serving oil refineries, coal movement, steel mills, chemical factories, and paper mills. It was so polluted by 1969 that it "had caught fire at least a dozen times since the Civil War and probably many more than that. By the 1930s, the local press was more critical of a lack of firefighting services to put out the river fires than they were of the river burning in the first place."[34] As Doyle explained, a 1969 fire got a lot of attention because it was covered in a very prominent issue of *Time* magazine: the one from August 1, 1969, that covered the return of the Apollo 11 astronauts from the moon *and* the scandal at Chappaquiddick Island where Mary Jo Kopechne was killed in a car wreck from which Ted Kennedy escaped and in which he had been the driver. Doyle noted, "People everywhere bought the magazine to see Ted Kennedy and the astronauts but then browsed their way to read about the burning river in Cleveland."

The attention to water quality brought forward by this burning river, the growing attention paid to water quality in other publications, and several other factors converged to lead to a flurry of environmental legislation in the following years: the Clean Water Act, the Clean Air Act, the Safe Drinking Water Act, and the National Environmental Policy Act. These policies led to the Environmental Protection Agency's (EPA's) permitting program, which had regulatory authority (a stick) and money to invest in programs (a carrot) to clean up water. Since then, in addition to the energy spent to clean up drinking water, energy has been invested to clean up wastewater before it is dumped into rivers.

## ENERGY FOR WATER

The idea that water is just out of reach is a common part of our language and culture. The old seaman's ditty "water, water everywhere, nor any drop to drink" from Samuel Taylor Coleridge's 1798 poem "The Rime of the Ancient Mariner" captures the essence of being surrounded by seawater that is unsuitable for consumption.

And it is no surprise that the word "tantalize" has its roots in a water-based legend. The Greek gods punish Tantalus, a son of Zeus, by giving him great thirst and forcing him to stand in a pool of water that always recedes as he leans down to take a drink. Such a myth feels like a fitting parable for humankind's relationship with abundant water resources that seem to be forever just beyond our reach. Because we use energy to bring that water within reach, we might say that energy lets us overcome the power of the Greek gods.

At the heart of the world's distribution of water is the global hydrologic cycle, which is large, powerful, and continuous. As described by the US Geological Survey, "Earth's water is always in movement . . . Water is always changing states between liquid, vapor, and ice, with these processes happening in the blink of an eye and over millions of years."[35] The cycle can change the intensity of flooding or drought in different places over time, but it never stops.

While the hydrologic cycle is global and abundant, it is also problematic. There is plenty of water in the world, but it is typically in the wrong form (salty), in the wrong place (far away, deep below ground, or in snowpack on top of mountains), or available at the wrong time of year (such as in parts of India, where there is plenty of water during the monsoon season but a shortage the other eleven months of the year) for human uses. We spend energy to get water to the right form (fresh, clean, treated, and heated), right place (our homes or businesses), and right time (when we need it). In many respects, the hydrologic cycle is an image of water abundance only as long as we have energy to transport water or transform it to our purposes.

Unfortunately, freshwater is a small fraction of the total: about 97.5 percent of the world's water is saline or brackish, and only 2.5 percent is freshwater.[36] Of the world's freshwater, much of it is locked up in ice or other locations that are hard to access. That means we have to spend energy to get water to where we want it. In particular, we need to spend a lot of energy lifting water from subsurface aquifers or lifting water into the atmosphere and then catching it on its way down.

The pump driving the water cycle is the incoming energy from the sun. Just over half of the earth's incoming solar radiation is consumed in the process of evaporating water.[37] Essentially, the sun raises water to a high altitude in the atmosphere, after which gravity brings it back down as snow and rain. If we could capture the entire gravitational potential of that elevated water, it would give us 13 terawatts of power, which is about twice the rate at which the entire globe consumes electricity. As it rolls back down to the oceans, we harness it for power, irrigation, drinking, and many other purposes. Then the whole cycle starts again. In other words, evaporated water in the atmosphere is a key driver of the water cycle.[38] There is approximately eight times more water stored in the atmosphere than in all of the world's rivers combined. There is 150 times more water in glaciers and snow than in all of our lakes combined. There is plenty of water.

The Earth also does some water lifting for us. There are places where the groundwater comes out naturally or under its own force. Those are called springs if the outlet is natural and flowing artesian wells if the outlet is man-made. Artesian wells are dug, drilled, or cut deep into the ground to a section of the water table that is under enough pressure that the water comes to the surface on its own. Prolific wells can produce more than 300 gallons per minute. Such wells allow us to avoid the nuisance of having to lower a bucket or use a hand pump to raise the water. Sometimes the pressure is so high that the water shoots out of the well several feet. Such an arrangement—with the Earth doing the pumping for us—is handy. And it's also safe, as it spares careless people or unattended children the risk of falling down a conventional open well.

Thankfully, the sun and the earth do a lot of the lifting for us. But wherever the benefits of their hard work aren't available, we have to do the lifting with other forms of energy, such as muscle power from humans and work animals, or mechanical power from electrically powered pumps.

Many different devices have been invented to aid in lifting or moving water: wheels with scoops, levers, buckets, pulleys and pails, treadmills or merry-go-rounds with horses or mules driving a lift system, and Archimedes' screw, first developed thousands of years ago. The ancient Egyptians widely used the latter, which manually turns a screw to elevate water. The tight coil of connected blades could raise water as long as it was continually operated. Its invention is attributed to Archimedes in the third century BCE, and it is still seen today in modern water-lifting

stations at amusement parks, water treatment plants, and elsewhere. It turns out that robust designs are still useful thousands
of years later.

Water needs so much energy for pumping because it is so
dense: it weighs 8.34 pounds per gallon. That density is one of
the reasons that water is so valuable as a coolant or heat carrier, and also means it takes a lot of energy—and therefore effort
or money—to bring that water uphill. The energy needed for
pumping water depends on how far the water needs to be raised,
the rate at which it is raised, pipe diameter, friction, and so forth.
The energy that is needed to raise water up out of a well is the
energy required to overcome the force of gravity, which wants
to pull the water back down. Raising a 2.75-gallon bucket of
water (about 10 liters) a distance of about 330 feet (100 meters)
requires approximately 10 BTUs (or 10,000 Newton-meters) of
energy. Pumping the water up only 33 feet (10 meters), for example by raising the water from a river to the top of a nearby
riverbank, requires 1 BTU (or 1,000 Newton-meters) of energy.
That doesn't seem like much, but for large volumes, the energy
adds up, and the distance water has to be elevated drives the
energy needs.

For continuous flows, it is more useful to look at the power
that is needed. Pumping 2.5 million gallons per day (29 gallons per second, or 110 liters per second)—enough water for
1,900 average Americans—out of an aquifer 330 feet below the
ground requires 107 kilowatts of pumping power. Keep in mind
that a typical house needs 1–3 kilowatts of power on average
to run the whole place, so a pump that size consumes the same
power as approximately thirty to one hundred homes.

A medium-sized US city with a million residents might need
about 150 million gallons of treated freshwater per day. Raising that water from a nearby river to elevated water treatment
plants over a height of 100 meters requires a little more than

6 megawatts of pumping power. Large wind turbines produce approximately 1 megawatt apiece, so that city would need a half-dozen of them running full-bore just for pumping water up from its source to the top of the hill, after which it can flow downhill to customers.

Imagine pumping that water by hand, the way it once was. In poorer parts of the world that reality still reigns. In Southeast Asia and sub-Saharan Africa, it is usually the responsibility of women or girls to fetch the water. Those women and girls miss hours of work or school each day to get water from far away, carrying the heavy jugs of water balanced on their heads or hanging from bars resting on their shoulders to cover a distance of over a mile between the well and the school. But once they have the water, they are not done, as the water needs to be treated before it can be drunk. In remote villages where piped water systems and centralized water treatment with electrical pumps and other advanced techniques are not available, water is treated the old-fashioned way: it must be boiled. There's a similar story for getting fuel: women often have to collect fuel from remote areas, and again they are vulnerable to kidnapping or assault along the way. In many parts of the world, women and girls spend 1.4 hours per day collecting fuelwood, which keeps them out of school.[39] Using modern energy to spare girls the effort of those laborious chores gives them a chance for more school-based education.

When they return after having fetched water and fuels, they need to use the fuels to boil the water. Those fuels—including crop residues, animal waste such as cow dung, wood, and un-treated coal—are burned in stoves used for cooking and heat-ing. Unfortunately, the dirty, inefficient cookstoves perform badly, producing indoor air pollution that has been linked to the premature death of about 2.5 million women and children every year. In other words, antiquated energy and water systems put women at risk when they are collecting the water and the fuels and when they are using the fuels to treat the water. These old energy and water systems literally deprive girls of their education and kill women by the millions.

Such archaic, labor-intensive approaches to energy and water take a toll on prosperity and economic opportunity.[40] In many places, the world's poorest women are also traditionally responsible for planting and harvesting crops, milling grain, and fulfilling household chores. These responsibilities leave little time for an education or employment outside the home. Even if they could go to school, they might not have the lights they need to read books at night to study or have to miss hours of school to fetch water. In turn, they have few or no opportunities to work, earn an income, and gain independence, which perpetuates poverty.

Although the scenario I described sounds like something from the developing world somewhere far away, it is also part of the American experience in the not-too-distant past. The University of Missouri in 1920 issued a poster as part of a campaign to encourage farm owners to modernize their water systems. It is aptly titled "The Farm Woman's Dream," and shows a woman—presumably poor and dressed in rough clothes or rags—carrying water without gloves from a hand-pumped well in the freezing cold along an icy path uphill. The dreary image evokes a sense of hard labor associated with getting water into our homes. In the upper part of the poster is the farm woman's purported dream: she is nicely dressed in short sleeves, comfortably indoors, opening spigots at a sink, with hot water coming out, its steam curling to the ceiling. The poster is targeted at women, not men, a telling example of who endured the greatest burden from not having access to modernized systems, and who would reap the greatest benefit from modern improvements. The burdened, vulnerable woman in Africa today, struggling to fetch water, is not much different than the burdened, vulnerable woman in the rural United States a century ago. Access to water and energy turns the story around.

For those women, their dream—the solution—is the same: a modern water system (with pipes and pumps) and a modern energy system, including electricity to drive the pumps and fuels to heat the water, can lessen the tedium of chores. Novel electrical appliances such as the dishwasher and washing machine provided women in the United States with even more freedom.[41] This point was not lost on US leaders and was one they used to differentiate American technological advancement from Cold War opponents. In the so-called kitchen debate of 1959, US vice president Richard M. Nixon and Soviet premier Nikita Khrushchev were touring an exhibit of American technology when Nixon stopped at a model American kitchen. Nixon highlighted the electric appliances—the refrigerator, range, and dishwasher—that "make easier the life of our housewives."[42] Chores like cooking and cleaning were no longer as time-consuming, complicated, or dangerous, which enabled women to pursue opportunities outside the home beyond menial labor. In turn, American culture began to shift to accommodate working women.

Energy for pumping water is just the start. We also need to get the water in the form we want: fresh, clean, safe, heated, treated, pressurized, frozen, or chilled. The amount of energy needed to treat water and wastewater to a suitable form depends on a variety of factors, such as how contaminated the source water is, the nature of the contamination, and what the water will be used for, plus the physical features and treatment approach of the facility. Because of the accumulated wastes from centuries of intensifying industrial activities and population growth, natural waterways became quite polluted. As pollution increased, the energy needed to treat water increased, too.[43] Cleaning dirty water is relevant to the myth of the unicorn. Part of the allure of the unicorn was that its horn was said to have a magical power that could rid a stream of poison.[44] Today we have replaced the magic of the elusive unicorn's horn with energy.

Despite the importance of clean water and sanitation, the close connections between public health and water supplies were not revealed scientifically until the mid-1800s. The first scientific identification that cholera is spread by water sources contaminated with human waste was made in 1849 by Dr. John Snow in London, who was able to determine that public wells that drew water from the heavily contaminated Thames (at the time, untreated sewage was emptied into the river) were the source of an outbreak of cholera in 1848. London suffered another similar outbreak in 1854.

The sudden spike in mortality exceeded that of London's famous plague episodes, producing spectacular consequences such as five hundred people dying in a small neighborhood over the span of just ten days. Unfortunately, Dr. Snow's findings that human waste was contaminating water and killing people were rejected by Parliament because they did not fit prevailing ideologies and because the actions that would be required to fix the problem were deemed too expensive. Similar rejection is offered for today's climate scientists, who tell us our waste is killing us, though in a much slower and less direct way, and that fixing the problem will require significant investments in new infrastructure. Dr. Snow was later vindicated as a hero, and perhaps the same fate awaits our present-day scientists who warn us of climate risks and offer solutions.

Parliament eventually acted, though it took an event known as the Great Stink in London in 1858 to spark the initiative. Although untreated sewage had been dumped into the Thames for decades, the currents had done a great service to the metropolis by washing the waste away to sea. However, the combination of heat wave and drought that summer meant there was less water to dispose of the wastes. The stagnant water and air produced a remarkably noxious aroma, triggering a temporary suspension of Parliament, whose building sits within smelling distance of the Thames. The event was unfortunate and unpleasant in many ways, but it led directly to the creation of an ambitious public works project to insert 1,200 miles of sewers into a crowded city of 3 million people. Doing so not only addressed the problem of waste disposal, but it also created the lovely river embankments that still stand today as a key piece of London's urban landscape along which many people stroll.

In the mid-1800s, Londoners could solve their water problems by simply flushing waste farther along the Thames. But today, with a higher global population and bigger cities, there is no "away." It's impossible to rely on dilution as the solution. Instead, industrialized societies invest energy: energy for water treatment and energy for wastewater treatment. And, as in the London experience where the sewers became walkways, if we do it the right way, we can solve our water problems while simultaneously building structures we can use for other purposes.

At around the same time, the city of Chicago was facing similar challenges. It drained its sewage into the Chicago River, which fed into Lake Michigan. That means Chicago was using the lake as both its toilet and its drinking water. Around the same time London was grappling with its own episodes, Chicago had outbreaks of cholera in 1849 and 1854, killing hundreds of people each time. Doyle summed it up by noting, "Quite simply, Chicago's water and waste disposal systems made it a dangerous place to live."[45] A few decades after Snow connected the dots between cholera and dirty drinking water, scientists in the United States scientifically confirmed the relationship between sewage-contaminated waters and typhoid fever, which ravaged cities.

As cities became bigger and had less river distance between them, the accumulating wastes became more dangerous. Eventually, cities built out their water systems to treat the incoming water to make it safer to drink. Better techniques for water treatment incorporated chemicals, blowers, stirrers, filters, and other devices—all of which required energy—to improve water quality. Ultimately, the fact that municipal water treatment systems enabled people to drink water without worrying about it can be considered one of the greatest public policy achievements of modern civilization in the last 150 years.[46] Similar changes will take place in the developing world when we work harder to promote greater access to electricity and the adoption of modern energy tools. Improved living conditions will help alleviate poverty among women while also giving them more choices in life.

Although the scientific and democratic advances since the industrial revolution have been significant, the largest public health problem globally remains the more than 1.1 billion people without access to clean water sources for drinking, cooking, and washing.[47] That number is expected to grow to 1.8 billion people by 2025. In China alone, 100 million people lack improved water sources, and 2.6 billion globally remain vulnerable to waterborne diseases because they lack access to sanitation (which is a polite word for wastewater treatment). Nearly 4.8 billion people, or 80 percent of the world's population in 2000, reside in areas with significant threats to water security or biodiversity. Improving water quality is a significant way to improve public health worldwide.

Indoor plumbing is one of the key luxuries that defines a higher quality of life. In the United States it is so widespread we take it for granted, but around the world people want for it. Delivering universal access to clean water to improve public health will require a lot of energy for treatment and transport to where it is needed. And if we're not careful, just throwing more energy at the problem might make other problems worse. Even though the energy can be used to clean up the water and improve public health, if we use energy that pollutes, then it might undo those benefits by depositing pollutants back into the water systems. Air laced with pollutants from smokestacks and tailpipes can cause premature mortality or weaken productivity because of employees taking sick days, either because they do not feel well or because they have to stay home to tend to a sick child. Emissions from our energy consumption accumulate in the air. With the right meteorological conditions, those emissions are converted into dangerous chemicals such as ozone, or fine particles that settle deep into our lungs and cross into our bloodstreams, or acids that get deposited through acid rain into waterways, killing ecosystems.

Beyond treating our water, energy also lets us heat our water, which is critical for sterilizing medical equipment, washing our hands, and cleaning scrapes and wounds. The idea of water as a purifying substance is propagated in a variety of religious traditions that include common features such as holy water or ritual bathing.

Dirtier water generally requires more energy for treatment, and end-uses that require high standards of cleanliness also need more energy. Hospitals, semiconductor cleanrooms, and food preparation facilities need water that is much cleaner than what is required for cooling industrial equipment or irrigating farms. That means there is great variability for the energy intensity of water treatment.

Coal has a dual personality with respect to water quality: coal mining and use are sources of water pollution, but coal can also help to clean water. Jeff Goodell, author of *Big Coal*, cleverly noted that coal mining can be done in two ways: you either remove the coal from the mountain, or remove the mountain from the coal.[48] That latter approach is known as mountain-top removal mining, and it is really impactful because the tops of mountains are removed and dumped into the neighboring valleys, burying waterways and polluting streams. Then when coal is burned, it releases pollutants and acid-forming gases that pollute the water further. Subsequently, the ash that's left behind can flood entire valleys. But, ironically, the same coal that dirties our air and water also makes electricity that we use to clean our water. And using an electric cookstove powered by electricity from a coal-fired power plant far away is cleaner than burning cow dung in our homes.

Bottled water, which is energy-intensive, also has a dual personality.[49] Energy is needed to process, bottle, seal, and refrigerate the water, but it turns out the energy to make the plastic bottle itself is the biggest piece in the product's life cycle. And if that bottle is moved by trucks, planes, or ships over long distances, the energy consumption goes even higher. In modern cities that have well-functioning piped water systems that draw from local sources, the water is produced in an energy-efficient and affordable way. By comparison, bottled water is one to two thousand times more energy-intensive per liter than municipal water, depending on how far the bottle is shipped.[50] In those situations, the extra money and energy for bottled water seems like a wasteful indulgence that offers little additional value. However, after a natural disaster such as a hurricane wipes out a local water system, or in developing countries where water systems are contaminated or do not exist at all, bottled water can be a lifesaver.

We also use energy in our homes for other creature comforts related to water. Sometimes to remove water from our hair with an electric hair dryer, to boil water to make tea with an electric teakettle, or to circulate water in backyard pools. Those devices are more energy-intensive than most people would anticipate. And their total power demand can strain the grid. In Venezuela, it is fashionable for women to style their hair, making hair dryers a common household appliance. But a hair dryer requires 1–2 kilowatts of power, which is about as demanding as a plugged-in electric vehicle (though the hair dryer might only be drawing current for a few minutes, whereas the vehicle might do so for hours). If millions of hair dryers operate at roughly the same time, it creates a surge in electrical demand that can measure in the gigawatts. Keep in mind that 1 nuclear power plant typically provides about 1 gigawatt of power. In early 2016, Venezuela was suffering a significant drought. Because the country gets a large fraction of its power from hydroelectric facilities, the drought raised the risk of power shortages. To help forestall a national crisis, President Nicolás Maduro "urged women to stop using hairdryers and offered alternative styling tips as the country's energy crisis continues." He even went on to say, "I think a woman looks better when she runs her fingers through her hair and lets it dry naturally."[51] That a president found it reasonable to give hairstyle suggestions to women is remarkable on many fronts, but that he did so for reasons related to energy reliability is striking.

If Venezuela is obsessed with styled hair, the English are obsessed with tea. And soap operas. *East Enders* is a very popular soap opera aired on the BBC. After the show ends, 1.75 million teakettles are turned on almost simultaneously. Each teakettle requires 1–2 kilowatts, similar to the hair dryer, creating a power surge of about 3 gigawatts for 3–5 minutes.[52] This surge is reminiscent of the wastewater surge that happens when Americans watch the Super Bowl and take nearly simultaneous toilet breaks during commercials. The English teakettle surge requires having standby power on the ready, including power plants in France and a pumped hydroelectric facility in Wales that stores electricity by elevating water so it can run downhill at a moment's notice when the power is needed.[53]

Pool pumps are another cause of electricity demand. Each one draws about 0.5 to 1 kilowatts, but unlike the teakettle or hair dryer, they work around the clock during swimming season to circulate water in backyard pools. One study concluded that pool pumps are responsible for nearly 8 percent of the peak power demand in a neighborhood, and can be 20 percent of a home's peak electricity demand in the summer months.[54] I laughed when I read this study because it used homes in Southern Ontario, Canada, as test cases for analysis. Given its northern climate, Canada obviously isn't a mecca for summer swimming. Doing a swimming pool study in Canada seems a little bit like doing a study on snowblowers in Houston, Texas.

When we add up all the different ways society uses energy for water pumping, treating, and conditioning, the tally is significant. In total, 4 percent of global electricity consumption is used to extract, distribute, and treat water or wastewater, along with 50 million tonnes of oil equivalent, mostly diesel for irrigation pumps and gas in desalination plants.[55] In the United States, we use about 13 percent of our total energy consumption for our water and steam systems.[56] About a third of that 13 percent, or about 4 percent, is to heat water to make steam for industry. Another third is just to heat water for our homes and businesses. We use more energy to heat water in the United States than countries like Switzerland use for an entire year for all purposes. In total, we use more energy for our water system than we use for all the light bulbs in America.

## SOLUTIONS

Given the interdependence of energy and water, there are some crosscutting benefits and some cascading challenges. The availability of water enables the availability of energy, and access to energy improves access to water. But because they rely on each other so much, it also means that a problem in one becomes a problem in the other. That means we need to think of ways to use the two resources to improve the reliability of both without exacerbating problems.

One way is with smarter technologies, using sensors, data, and cheap computing to optimize how we monitor and manage our systems. Consumers in their households should have a clearer idea of how much energy and water they are using from different appliances and usage patterns. Another way is to use advanced technologies. Israel, which is in a water-stressed area and for national security reasons cannot rely on its neighbors for a water supply, has turned to a combination of conservation (to reduce its demand for water) and advanced technologies such as robust desalination systems (to increase its supply). The Israeli approach worked, and now its technologies are being exported around the world.

Water recycling is also an option. Some pejoratively refer to this approach as "toilet to tap," but it works. Like Israel, Singapore does not feel it can rely on its neighbors in the long run for its water. Therefore, it built a water recycling facility in 2000, calling it NEWater. The system provides 60 million gallons per day, or 30 percent of the drinking water, from reclaimed wastewater with the goal of tripling its capacity by 2060. Water recycling and toilet-to-tap treatment systems are also valuable for the military. Shipping water to the front lines is an expensive and deadly proposition. The long supply convoys span thousands of miles and are soft targets for enemies, meaning that the cost of water per gallon ultimately is several orders of magnitude higher if delivered to the military theater than at the local grocery store or from our taps. Consequently, the US military spends a great deal of money testing and deploying on-site water treatment systems that can use energy to make potable water from degraded streams. These systems cost a lot of money and require extra space to be shipped to the forward operating base, but after that, they spare the need for a lot of shipments of water, saving lives.

The International Space Station also has a reclaimed water system to produce drinking water because shipping water to space is exorbitantly expensive.[57] It costs from $10,000 to $90,000 per pound to ship cargo to space. That means shipping fresh water to astronauts in space costs anywhere from just under $100,000 to nearly $750,000 per gallon. Because of the high costs for freshwater, the space station collects and treats the gray water from washing, urine, and condensed moisture from breath and sweat to be drunk again. The station does not have any black water because the astronauts do not use flush toilets.

Interestingly enough, my research at Stanford University for my PhD was a part of a project at the National Aeronautics and Space Administration (NASA) Johnson Space Center to create an onboard water treatment system for the space station. As a graduate researcher in mechanical engineering, I invented and deployed laser-based sensors that could measure a variety of trace gases. One of my patents from that work was for a sensor that would measure very small concentrations of ammonia, which was useful for monitoring the health of the water treatment system.[58]

Other approaches include switching to energy-lean water systems, and water-lean energy systems. Renewable electricity technologies such as wind turbines and solar panels do not use heat to make electricity, so they do not need cooling water. They need small volumes of water for manufacturing components at the steel mill where the turbine parts are fabricated or the semiconductor factory where the solar panels are printed, and they also use water for cleaning equipment in the field. Other than that, their water needs are minimal. In parallel, water treatment systems become more efficient with time, saving energy while improving water quality. Integrating low-carbon sources of energy such as wind or solar with water treatment systems that perpetually improve offers multiple benefits.

There are also some novel ideas involving passive systems for cleaning water. One is the LifeStraw family of devices manufactured by Vestergaard, a Swiss company. They range in size from a personal straw or water bottle up through family-sized tanks to serve a household. Their treatment technology solves some key water treatment needs—removing bacteria and parasites—without active energy inputs.

The good news is that conservation is a powerful approach. And because of their interdependence, saving water saves energy while saving energy saves water. Thankfully, there are many cost-effective ways to conserve energy or water in our homes.[59] A study by Professor Ashlynn Stillwell's team at the University of Illinois at Urbana-Champaign concluded that a typical house in the United States could reduce its electricity consumption and water consumption by 7,600 kilowatt-hours and 39,000 gallons annually in a cost-effective way; that is, the cost of implementing the conservation and efficiency methods pay for themselves from reduced utility bills. These retrofits include the mundane, such as more efficient light bulbs and low-flow toilets, as well as sophisticated appliances and smart controls. This lesson about conservation as a useful approach will be a recurring theme in terms of how we can save resources and make our cities and transportation more efficient to save money.

Our modern energy system needs water, and our water system needs energy. Ultimately, energy and water are so tightly linked, you can think of each one as synonymous with the other. Unfortunately, that means the risks of failure in one system can quickly cascade to the other. And these two—energy and water—are the underpinnings for the subsequent factors of society that matter: food, transportation, wealth, cities, and security.

Chapter 2

# FOOD

Food is a form of energy that our bodies need to function. And the food system—from farm to fork—requires vast amounts of energy as inputs to function. Just as modern forms of energy revolutionized our access to water through the use of electric pumps and treatment systems, the food system was revolutionized by the arrival of modern energy forms such as diesel for tractors, oil and gas for agrichemicals such as fertilizers and insecticides, and electricity for refrigeration. Advanced energy liberated us from the shackles of manual labor and reduced the vagaries of nature by enabling mechanization, fertilization, and refrigeration while slowing the impact of infestation. The consequence was a green revolution in the twentieth century that transformed agriculture and caused farm productivity to grow exponentially. It is a remarkable story of progress that so few people can feed so many and that energy has steadied the growth in acres of agricultural land despite an explosion in population. The good news is that developed countries with sufficient energy went from food scarcity to food abundance. The bad news is that there are still more than 1 billion people who are undernourished or food insecure,[*] and richer countries now have a food waste problem because of all the excess.

A typical person needs 2,000 nutritional calories per day just to stay alive. That's about the same as 10,000 BTUs of energy per person per day. Over the course of a year, that means 1 quadrillion BTUs of energy each year are needed to feed 325 million Americans, and 25 quadrillion BTUs are needed to feed the world's total population. Overall, the United States consumes 100 quadrillion BTUs (or quads, for short) of energy, and the world consumes 600 quads. That means we consume much more energy than is needed to feed ourselves; the rest of that energy is for other purposes, such as transportation, industry, entertainment, and so forth. It also means we have plenty of energy to solve the global crisis of food insecurity, but other barriers—politics, corruption, market failures—keep us from getting food into the hands of the hungry.

It takes a lot of energy to run the entire food system, often at a ratio of about 10:1. The American food system requires about 10 quads of energy to grow, irrigate, fertilize, harvest, process, wrap, store, refrigerate, distribute, prepare, and dispose of the 1 quad of food energy we need to thrive. It's not just calories we need to survive, though—we need a mix of nutrition and calories. We cannot live on lettuce alone, for example. The good

---

* According to the Food and Agriculture Organization (FAO) of the United Nations, food insecurity is "a situation that exists when people lack secure access to sufficient amounts of safe and nutritious food for normal growth and development and an active and healthy life." See Marion Napoli, "Towards a Food Insecurity Multidimensional Index (FIMI)" (master's thesis, Roma Tre Universita Degli Studi, 2011). The way I think about it, to be food insecure means that a person is within twenty-four hours of going hungry, and that minor events—a storm that interrupts food delivery systems or a hiccup in delivery of the weekly paycheck—means that person will not be able to eat enough food or a sufficiently varied diet.

news is that we have the energy available to meet these needs and that energy abundance enables food abundance. Making food widely available is a critical first step to forming advanced societies with stable institutions and the rule of law, but along the way the bad news is that the foodprint—the energy, water, land, and emissions footprint of food—is significant.[1] Allowing food to spoil and throwing away perfectly edible foods exacerbates the situation because we waste much of the energy, land, and water we invest while much of the world goes hungry.[2] Solving this combination of food problems—increasing food access while reducing its impacts—remains one of the world's most vexing challenges.

## ENERGY FOR WATER FOR FOOD

As you may have guessed, the story of food and energy begins with water. Photosynthesis, which ultimately is the conversion of solar energy into chemical energy stored in plants, lies at the heart of our food system. Photosynthesis needs a few key ingredients: sunshine, carbon dioxide, and water. And water is often the limiting factor, as $CO_2$ and sunshine are available globally, though perhaps irregularly. Nutrients such as nitrogen (N), phosphorus (P), and potassium (K) also play a critical role in the metabolism of plants.

Moving water for field irrigation has been a part of the story of civilization for thousands of years.[3] Long ditches and furrows would move water to enable concentrated crop irrigation. Once crops were grown, animals could be domesticated as a source of protein or for work. Oxen and horses would haul goods or pull plows, cows and goats would provide milk and meat, cats would control pests such as rats, and dogs would provide security. Managing water resources was the key enabler for increased food production, and it helped shift diets from those that could be supplied by hunting and gathering to other diets that allowed for civilizations to take root in one place and grow. That has been true for millennia. But since the twentieth century our food system has become dependent on newer energy inputs, such as petroleum, natural gas, and electricity.

As water tables fell or as nearby sources were degraded or depleted, more energy was required to get water that was needed to the fields and villages. Innovative mechanical devices such as Archimedes' screw allowed water to be raised from low-lying bodies of water into nearby fields, but required energy to operate. At some point the depths from which the water needed to be raised, the distances water needed to be moved, and the sheer volume of the water exceeded the ability of human or animal pumping to meet the needs. Modern energy systems in the form of steam engines and eventually electric pumps met this challenge. With the ability to raise water from depths of hundreds of feet and move it along the surface thousands of miles, we could make deserts bloom, opening up the prospect of growing fruits and nuts in semiarid locations such as California's Central Valley.

Reliable water was a key pathway to food abundance, which enabled stable societies. Stabilized societies could take on multigenerational challenges. It has been posited that because of increased and stabilized agricultural yields in Europe, leaders could decide to launch ambitious multigenerational projects, such as the construction of cathedrals over hundreds of years.[4] Many of those remarkable buildings still stand today as testaments to the powerful and lasting effect of food abundance.

As societies got richer and agricultural productivity improved, their citizens switched to diets with more protein, moving from tubers, seeds, and nuts to animal proteins. This improvement in human diets helped people grow stronger and live longer. But there were also downsides. Animal proteins are notoriously land-, water-, and energy-intensive because of all the inputs, including fertilizers and irrigation, needed to grow the feed for their final stages of fattening. Despite the needs at the feedlot, by one estimate, 80 percent of cattle feed is not edible for humans, so cows are also upcycling low-grade materials into higher-quality foods.[5] I often joke in my class lectures that someday we need to invent a portable, flexible bioreactor that can take low-grade biomaterials like inedible grasses and convert them into compact, dense, high-value proteins. I then follow with the punch line that we could call this novel technology "a cow."

In all, on top of the story of innovation and progress, the food system has a much bigger environmental footprint than most people expect, and we need to take a view of the whole system to identify the right kinds of solutions.

## ENERGY ON THE FARM: FERTILIZATION AND MECHANIZATION

Water availability is just one piece of successful agricultural systems. Nutrients are also critical. One of the most important is nitrogen (N), which is a primary component of amino acids that are building blocks for plant protein and ultimately for animal protein. Thankfully, nitrogen is abundant as the most common constituent of the atmosphere in gaseous, diatomic ($N_2$) form. But many crops need liquid or solid nitrogen in their roots to be effective.

Over time, nature has done its part: through a variety of processes that have unfolded for millennia, fertile soils are rich in nitrogen. However, as crops were planted in successive years, the nitrogen was depleted from soils as it was embedded in the crop products, causing some fields to lose their productivity. The development of three-crop rotation helped mitigate that challenge. One of the crops put into rotation is a legume that fixes nitrogen.* By rotating crops in a specific way, nitrogen would be depleted from the soil with one crop and replenished with another.

Other natural fertilizers include manure and urine, which are rich in nitrogen and phosphorus. In agricultural operations that mixed different crops while also raising animals like cows, goats, pigs, and chickens, the animals would produce the fertilizer to grow their own feed while also ultimately serving as food for the farmers and their customers. This system, while robust, having lasted millennia, was inherently limited in scale, as it was impractical to spread manure beyond a certain radius from the farm. Guano, or excrement from birds and bats, was also a common commercially available fertilizer, but natural reserves of it were being depleted.

That is where modern energy changed agriculture forever. Nutrients get naturally restored into soil very slowly. But during the early part of the twentieth century, German chemists developed the Haber-Bosch process. This series of chemical reactions converts natural gas, primarily composed of methane ($CH_4$), into ammonia ($NH_3$), which can be applied in liquefied form to fields as a rich source of nitrogen for fertilizing the crops. This process was inherently transformative for agriculture because it enabled much higher yields per acre and per farmer through the chemical inputs. Farmers were no longer limited by the supply of manure for fertilization, and crops grew much more rapidly and robustly. After World War II there were worldwide food shortages because of the damage the wars caused to agricultural fields. In addition, the war had killed so many people that there

---

* "Fixing nitrogen" means putting it into a form that plants or other living organisms can use.

was a shortage of farmers. Fertilizers synthesized from fossil fuels overcame those limits, helping to improve food abundance and access for hundreds of millions of people.

The adoption of synthetic fertilizers was rapid: "By the 1960s, over 90 percent of all cornfields in the United States were fertilized with synthetic nitrogen. The amount used per acre increased by a factor of 10 over the second half of the twentieth century, when farmers began applying more than was needed to ensure maximum crop yields. As farmers used more and more fertilizer, the plants were unable to absorb it all—and more of the nitrogen ran off the fields, into the irrigation ditches, and on to streams and rivers."[6] The use of fertilizers, like any energy advance, increased crop productivity but came with its own downsides. The runoff from fields created environmental problems as the fertilizers triggered the growth of algae in bodies of water. As the algae died, bacteria feeding on them used up the oxygen in the water, creating dead zones.

Fertilizers were just one part of the story. Though their use is controversial, there are a variety of other agricultural chemicals, such as herbicides, pesticides, and fungicides, all made from petrochemicals, that reduced crop loss significantly. Plagues of locusts or other pests could wipe out crops across a wide area in very little time, doing untold damage and putting food for large populations at risk. Energy-derived pesticides helped prevent this kind of catastrophe.

There is no doubt that the application of these chemicals has transformed agriculture. Most of these transformations are positive—food scarcity is less of a risk, crops are more resilient to nature's challenges, and productivity is significantly higher—meaning more food is produced on less land, which has environmental benefits. But the risks of those chemicals making their way onto and into our foods from spraying and our ecosystems through runoff leave consumers with nagging doubts about the ultimate value of the trade-offs as the environmental damage accumulates.

Agriculture requires a lot of work: plowing, harvesting, and processing crops. For most of human history, muscle power performed this work. Small farming operations would use hand tools and human muscles. Locations with domesticated animals would use beasts of burden, such as horses or oxen, to pull plows and do the heavy lifting. And a significant portion of the food that was grown would be dedicated as feed for those animals, leaving fewer bushels of remaining crops as product to be sold at market. But the application of modern energy tilted that balance in our favor.

Agriculture also requires tools. Fossil fuels—namely coal—enabled the production of more abundant, affordable, and durable metals by melting ores so they could be cast and providing a source of carbon for hardening the steel. These better metals in turn enabled the forging of more affordable and better tools—stronger plows, rakes, hoes, spades, and so forth—and these tools improved agricultural operations. Replacing older moldboards or wooden plows with metal plows improved farming because the stronger materials would last longer and could turn more dirt, helping to bring critical nutrients to the surface.

The advent of the internal combustion engine changed things again. Tractors, combines, and trucks powered by gasoline or diesel engines became available to farmers. They could cover much more ground more quickly on a tank of fuel than with multiple animals. That reduced the labor requirements, increased yields, and reduced the amount of feed needed for their herds. That the original unit of power for engines is the "horsepower" is a hat tip toward the idea that mechanized equipment such as tractors would replace horses in the fields, and potential customers needed to know how many horses would no longer be needed. The use of equids on US farms peaked at more than 25 million around 1920, but with the arrival of tractors their use plummeted steeply to just a few million on farms nationwide by 1960.[7] The shift from muscle power to diesel power was swift.

The movement from labor-intensive to energy-intensive farming moved the burden from human or animal muscles to fossil fuels. To produce 1 hectare of corn or other grains requires over 1,000 hours of human labor in nonindustrial parts of the world and about 10 hours of human labor in the United States and industrialized areas.[8] That means we have reduced the labor requirements to one-hundredth of what they would be otherwise. One human can do the work of about 100 other humans with energy resources at his or her disposal making up the difference.

## ENERGY FOR PROCESSING, PACKAGING, PRESERVING, AND PREPARING FOOD

Increasing agricultural productivity was only one part of the whole energy-intensive supply chain. Processing, packaging, preserving, and preparing the food required even more energy inputs.

Many raw agricultural products are consumed directly or simply packaged, such as eggs, fresh fruit, or vegetables. Other agricultural goods are processed further or converted into refined products of some sort, which might require milling, chopping, wrapping, precooking, canning, and so forth. Animals go to slaughterhouses where they are killed, skinned or plucked, and processed into cuts of meat. All of these steps require mechanical work or other inputs to achieve. Products like presliced bread require more energy inputs in the food system to get to that point.

For millennia, water power provided the necessary force for milling grains, as classic images of the medieval overshot waterwheel or Dutch windmills demonstrate. This need for water power gave rise to industrial city centers like Minneapolis, Minnesota, where major food companies such as Pillsbury and General Mills located themselves: close to the grains from the great plains of the United States, but at a point where water power was steady and available along the Mississippi River. Today the waterfront—with a prominent museum in an old mill—is a monument to that milling history. At its peak, from around 1880 to 1930, Minneapolis was the flour milling capital of the world.[9]

Then after all of that processing, additional energy is embedded in the packaging and containers. Glass milk containers, plastic bottles, aluminum cans, Styrofoam trays, and paper boxes all require energy for manufacturing and recycling. When I was a young boy in the early 1970s, we enjoyed the dying tradition of receiving fresh milk deliveries to our home twice weekly. The milk was delivered in reusable glass containers. When we were done, we would set the empty containers on the porch for the milkman to take with him after making a fresh delivery. That half-gallon, reusable glass milk container has about 4,500 kilocalories of energy embedded in the manufacturing of the bottle itself.[10] The half gallon of whole milk contained inside has less than 1,300 kilocalories of energy.[11] That means the container has more than three times as much energy in its materials as the liquid it carries. Scale that up across all types of packaging at a national level, and the embedded energy is significant, which shows how valuable it is to reuse materials whenever possible.

Preservation is another energy-intensive part of food's supply chain. Today we use electric refrigeration and freezing to preserve food. But in prior eras, we used salt, smoke, and ice. In locations where ice was available, it would be harvested in the winter and stored in cold locations such as cellars or insulated warehouses, and then distributed to people's homes, where it would be placed in the bottom of their food storage cabinet, giving us the term "icebox." This industry's prominence is still captured culturally, for example, in the opening scene of the blockbuster hit *Frozen*, which shows characters cutting and extracting ice to sell. Louis XIV, the French king, had ice pits installed at Versailles so that he could serve ice cream to his guests year-round.[12] In Michigan, icehouses lined the rail networks connecting Chicago slaughterhouses and meatpackers to eastern markets such as New York and Boston.[13]

Those Chicago slaughterhouses were transformative for the ranchers of the Great Plains and Midwest and the meat eaters in major population centers. Noted historian William Cronon described "Porkopolis," which was a crude name for Chicago's role in the meat industry: "For cattle that had grown fat on the grasses of the High Plains and the corn of the Iowa feedlots, Chicago was the end of the line. It was the place, more than any other, where animals went to die. In the grimy brick buildings that sprang up beside the great stockyard, death itself took a new form."[14] Chicago ultimately became a meatpacking center for cattle. But its origins were with hogs, because before the ready availability of ice harvesting, salting and smoking were the primary means of meat preservation, and both of those worked better with pork than beef to produce goods like salted bacon or hams and smoked sausages.

By building out an ice infrastructure for preserving meats, the farmers could liberate themselves from seasonal fluctuation. Pork was considered a winter meat because its production was isolated to the colder months. But in a warehouse where ice was used to keep temperatures low, pork could be processed in any season, turning it into a year-round meat. This is not unlike the practices of modern urban farmers or those with greenhouses who seek to control the climate and weather for their crops. And it is similar to the growth of the global food supply chain that gives us fresh fruits and vegetables no matter the season. For those Chicago-based meatpackers, rail-shipped ice was revolutionary. It also created new demand for ice: "local traders started to cut ice from the Chicago River during the 1840s, and had called nearby ponds into service by the end of the 1850s."[15] Ice ponds were becoming depleted. So, just as oil saved the whales and coal saved the forests, electric refrigeration saved the ice ponds. Once packers created a market for ice-packed meats, that evolved naturally into a market for mechanically refrigerated meats. They then moved on to other commodities for which chilled warehouses would provide a competitive advantage, such as chilled fruits.

The packers were efficient. They tirelessly sought to use every single part of the animals they dismembered, creating a host of goods including buttons, fertilizer, and glue. Chicagoans made the boast so frequently that it became a cliché: "the packers used everything in the hog except the squeal." This mentality was passed on to refiners in the oil industry, who despite their wasteful practices in the industry's infancy in the late 1800s ultimately evolved to the point at which they would use every drop of crude oil rather than discard it, an approach still in use today.

Getting people to eat ice-packed meats took a cultural shift. Consumers had learned the hard way over the natural course of their lives that eating spoiled meat was at best distasteful and at worst a major health risk. Because of their fears about the risks of consuming spoiled meat, people preferred local, fresh meats from a butcher rather than less-than-fresh cuts from a remote meatpacker. That tension between local versus remote foods remains a common element of local food movements doing battle with far-flung manufacturers even to this day. Energy is a big piece of that tension because without it, the entire concept of remote foods would be practically irrelevant. While those fears had to be overcome by meat sellers, the distance also had its advantages: that the packing plants were located in Chicago gave customers in Boston and elsewhere a comforting distance from the act of killing.

An older and more common approach to food preservation was the use of salt. Salt was so important to protecting meats and cheese from spoiling, it was considered a strategic mineral. Just as countries would go to war over oil in the twentieth century, in older times, countries would go to war over salt because it was so central to their food systems.[16] Just as oil was a motivator for wildcatters to explore foreign lands and put their lives at risk with the hope of getting rich, salt was an enabler of French explorers who traversed the ocean to collect fish they could sell back home. Their use of salt kept the fish from spoiling, giving them a critical advantage.

It's hard to imagine today how valuable salt was in the 1800s. When Rockefeller Center was built in midtown Manhattan in the 1930s, a custom art piece created by Carl Paul Jennewein and titled *Industries of the British Empire* was installed on the eastern façade of the building, above the entrance at 620 Fifth Avenue. This large bronze display—18 feet high and 11 feet wide—includes gilded figures that symbolize the nine most valuable industries for the British Empire. Salt is represented alongside other critical commodities such as coal and tobacco.[17]

We stopped going to war over salt when something better came along: electric refrigeration. By using electricity to run a compressor that would squeeze and circulate a refrigerant, enclosed containers such as an icebox or a house could be cooled, giving us refrigerators and air conditioning. The rise of electric refrigeration was critical because it meant food preservation could take place in hotter climates farther south and west, despite being far from ice-harvesting locations in the north. And electric refrigeration—which is flavorless—tastes much better than salt. Its invention transformed society in many ways. Kitchens were redesigned to accommodate this central appliance. Refrigerators quickly moved from being a luxury good to a necessity, and along the way they continued to grow in size. Refrigerators became such a big electricity consumer overall that they became a target of energy efficiency policies in the 1970s and 1980s. Despite growing in size from about 18 to 23 cubic feet on average, energy consumption dropped for refrigerators from an average of 2,000 kilowatt-hours to 500 kilowatt-hours per year as a result of a mix of policies and technical advances.[18] In 2015, refrigeration was responsible for 7 percent of residential electricity consumption in the United States.[19]

Electric refrigeration changed how people bought their food. Instead of shopping every day for fresh fruits, vegetables, and meats, consumers could go to a store less frequently and stock up with a week's worth of supplies. This new rhythm of less frequent shopping was a desirable modern convenience. The arrival of the refrigerator at the same time as the personal automobile coupled nicely with the rise of larger grocery stores. Grocery stores also created entire aisles of refrigerated and frozen goods for sale. Some of those freezers, lined up horizontally, are open at the top. At first blush that looks like a horrible way to be energy efficient, but cold air tends to sink, so the coolest air hovers inside the open freezers. The opposite approach was used for household appliances that combined a refrigerator and a freezer. Inexplicably, early versions put the freezer at the top, which is silly, given how heat rises. Because heat from the room and from the compressor in the refrigerator itself rose to the freezer portion, it had to work even harder to keep its contents frozen. More efficient designs later on moved the freezer to the bottom of the device, where it belongs, though amazingly, some refrigerators are still sold with the freezer on top.

All the energy inputs so far do not include transportation and only take us from the farm to the factory and then to our freezers and fridges. It takes even more energy to prepare the food that makes it to your plate. An electric oven or stovetop can require several kilowatts of power, which is about the same as what an air conditioner needs to cool an entire house or what an electric car needs for charging. The energy spent boiling water for pasta, baking, and roasting ends up being about 1 percent of national energy consumption, not including the energy to cool or heat the kitchen where this food preparation takes place.[20] Plus there is the energy embedded in the glass cooking dishes, stainless steel pans, metal knives, and plastic spatulas.

Modern homes might have stovetops heated by natural gas or electric induction; traditional cooking throughout the world and over millennia has been and remains much more primitive. In most places around the world, it is the woman's responsibility to prepare meals for the family, just as women are often tasked with fetching water. That means women are at risk of assault, kidnapping, or worse while collecting fuelwood. They then come home to use those solid fuels—if not fuelwood, then perhaps charcoal, cow dung, or straw—in simple cookstoves that generate significant amounts of soot, smoke, and ash. The indoor air pollution produced by these systems kills 2.5 million women and children prematurely each year.

Getting women access to advanced forms of energy and better cookstoves would prevent premature deaths from the smoke. Using modern energy to reduce the need for those laborious chores gives girls a chance to go to school and grown women other ways to be paid for how they participate in the economy.

One of the US founding fathers, Benjamin Franklin, also concerned himself with these problems. His Franklin stove was considered a breakthrough because it captured the smoke and soot, releasing them through a chimney rather than into the heart of the room. The Franklin stove also made the kitchen safer. Fireplaces and open hearths were a risk to women, not just from the air pollution they breathed but also because women's skirts and petticoats could catch fire. During colonial times, the second leading cause of death for women, after childbirth, was burns.[21] In addition to mitigating that risk somewhat, the Franklin stove, intended to be located in a prominent setting in a main room, also provided room heating for the comfort of its inhabitants.

## ENERGY FOR FOOD DISTRIBUTION

Food has been traded for millennia. The hunt for spices was one motivation for Marco Polo's trek from Europe through Asia. The quest for fish drew French explorers to the northwest Atlantic. There are early signs that Romans traded wheat from one province to another. Though globalized food trading has been around for specialized items for a long time, the widespread movement of so much tonnage is a relatively new phenomenon, starting after World War II.

Moving agricultural goods and foods around is also energy-intensive. Because of the rise of fossil-fueled trains, ships, trucks, planes, and cars, it became easy to create a sprawling global food system. Now it's possible to have 5,000-mile salads and fruits or vegetables that are in season year-round. This global supply chain brings us citrus fruits any time of the year because they're always in season somewhere. Our modern dishes might bring together fruits and lettuce from South America, nuts from North America, spices from China, and meats from Europe, all just to meet our preferences. And energy-enabled transportation is so swift and cheap that we don't even notice the distance the food has traveled. While this variety in foods and the global reach of the supply chain can have nutritional benefits and keeps menus from going stale, it also detaches eaters—all of us—from the places where foods are grown and can cause us to forget that food has seasons.

The energy requirements to move food around the globe can be enormous. But before the food supply chain expanded to a global network, food production first moved from our backyard gardens and small family farms to a rural hinterland providing for urban centers. This relationship fueled the growth of Chicago as a center of commerce, which Cronon billed "Nature's Metropolis." Animals from several states away within a radius of hundreds of miles would be shipped by rail to Chicago slaughterhouses for processing and packing before being sold to eastern markets. Raw grains like wheat and corn would make a similar trek, coming to Chicago for conversion into a standardized commodity, whether cornmeal or wheat flour, to be packaged and sold. As cities grew, the amount of land encircling them that was necessary to feed the urban dwellers grew in commensurate fashion. But eventually, the reach of our food system extended beyond the limits of transport by rail.

This global supply chain for food gives rise to the concept of "food miles"—how far our food has traveled to get to our plates. These large global supply chains fit naturally with global agricultural corporations, so the rise of food miles has occurred hand in hand with consolidation of smaller family farms into larger corporate structures. These corporate entities invest in energy and technology to improve productivity and yields on farms, which improves the amount of food available per person globally. That is both good and bad. Some of these companies have forever changed the shape of industrialized farming with controversial practices, such as seed licensing, that prevent family farms from deploying some traditional practices, such as seed washing, that have been used for generations. It also changes the character of farming from the cultural view of "American Gothic" with hardscrabble families working by hand to pass on an agricultural tradition. Instead, the image of modern farms is that of outdoor factories, with anonymous farmhands and machines working nonstop to produce large volumes of identical food. The corporatization of farming raises eyebrows because of concerns that an important cultural element in society is being lost. While it has been good for food abundance, it also helped separate people from food production. When asked where their food comes from, many will answer "the store," unaware of the extensive effort that was required to get the food to that location and condition.

Different forms of local food movements have sprung up around the world to reacquaint people with their food. In the United States, farmer's markets are growing in popularity, bringing consumers into contact with their food producers. In France, farmers rioted against imported foods that they felt put their livelihoods at risk. But the energy trade-offs of local food are not obvious. Sometimes the food grown remotely and transported half a world away requires *less* energy.

Take the example of lamb to be eaten in England. Typical sources for lamb in London are from the English countryside or from New Zealand, more than 10,000 miles away. While the instinctive response is that surely the local lamb would require less energy, in fact the lamb from New Zealand is a less energy-intensive option.[22] Shipping lamb over such a long distance includes a significant energy requirement, but oceangoing transportation is the most efficient way to move freight. Plus, there are many other energy inputs to growing lamb. Though local lamb does not have very far to go, its movement is exclusively by truck. The imported lamb from New Zealand will also move by truck for the last few miles of delivery, but in some cases the distance from the port might be shorter than from the farm. More importantly, the lambs in New Zealand are grass-fed in a temperate climate, and that grass grows naturally, irrigated by rain. The lambs in England are fed grains that themselves are water- and energy-intensive to grow because of the irrigation and fertilizers used. Those inputs overwhelm the life-cycle energy requirements for the meat. As a consequence, even though the lamb from New Zealand is a world away, it requires less energy overall.

There is another famous example with wine. A wine connoisseur in Chicago might contemplate whether it is better environmentally for her fine wine to be a California label air-shipped straight to her door or a French brand bought through a store. Both regions produce world-class wines. The wine in California is closer, but if it is air-shipped and the French wine is moved by container ship and then train and truck, it will be less energy-intensive to drink the wine from overseas than the domestic version.[23] So distance is not the only factor that matters.

This means the local food movement isn't necessarily better from an energy perspective. Food is better for the environment if it is locally appropriate to grow. If it is not locally appropriate, then the fertilizers and irrigation that make up the difference can be more energy-intensive. Some have tried to grow tomatoes and wheat in the deserts of Saudi Arabia and the United Arab Emirates. Growing food locally in these arid climates might be a rational effort to enhance food security—based on the fear that food imported from other countries could be cut off in a time of conflict—but the resource impacts are extraordinary. The amount of water needed to grow wheat in the desert is significantly higher than in more traditional growing areas like the great plains in the central United States. And if the closer food (e.g., California wine) is shipped in an energy-intensive way (e.g., planes versus trains), then the energy benefits of proximity are eradicated.

Despite the energy trade-offs, locally grown foods have other economic or philosophical benefits. My wife and I shop at farmer's markets because we like circulating our money in the local economy and we find philosophical and personal satisfaction from being connected to the people who grow our food. But if we are asking local farmers to grow foods that are not native to our climate, then we might be accidentally driving them toward higher use of energy inputs to get the crops to grow in local conditions. While we might save energy on transportation, the other energy inputs could swamp those savings.

## TOTAL ENERGY FOR FOOD

Because of these inputs for growing, packaging, refrigerating, preparing, and transporting food, the food system's relationship with energy is more extensive than typically anticipated. That energy consumption, combined with land use changes and emissions of methane ($CH_4$) and nitrous oxide ($N_2O$) from manure and fertilizer management on farms, makes the food system the second-largest source of greenhouse gas emissions in the world, behind the emissions from heat and power production and ahead of industry and transportation.[24] In total, the energy consumption for the food system in the United States is nearly 10 percent of annual national consumption. That means energy embedded in food is big enough to matter and warrants attention. Energy consumption for food is about two to three times bigger than all the energy consumed for lighting nationwide.[25] Think of all the innovations in light bulb design to reduce their costs and improve brightness over time. Innovation to improve the efficiency of the food system is still lacking, because it has not traditionally been considered a starting point for energy savings by policymakers, who have instead focused on light bulb and appliance standards and fuel economy requirements for automobiles.

In the end, an extensive analysis by researchers at Carnegie Mellon University found that changing what we eat has a much bigger impact than choosing where it is grown.[26] In particular, because meat is so energy-, water- and land-intensive to grow compared with grains, reducing the amount we consume would spare resources.[27] Red meat has 150 percent more greenhouse gas emissions than chicken or fish, so returning to the Catholic tradition of eating fish on Fridays or starting a new tradition akin to meatless Mondays could dramatically shift the resource requirements for our food consumption.[28] That means moderation—reducing (but not necessarily eliminating) consumption of meat—is a reasonable place to begin to save energy, land, and water while avoiding significant greenhouse gas emissions. Because our food system is so highly integrated, reducing meat consumption to zero might have important unintended consequences. For example, the meat system provides manure that is a critical input for growing other crops—notably grains, fruits, and vegetables—and uses land that could not be used for crops. If meat production ceased to exist, fertilizers would need to be manufactured by some other means, including from natural gas. That means everyone converting to vegan or vegetarian diets does not necessarily solve our food challenges.

## FOOD ABUNDANCE, OVEREATING, AND FOOD WASTE

We can also reduce the energy requirements for our food system by reducing how much we serve ourselves. In the United States we are simultaneously overfed and undernourished. We consume far more calories than we need to survive, yet our diets aren't healthy: we eat too many of the wrong calories (such as sugar and processed carbohydrates) and not enough of the right calories (such as those from fruits, vegetables, and healthy fats).

Worse, despite all this abundance, the United States is dotted with food deserts, locations such as inner cities where fresh, high-quality meat and produce simply aren't available. The only items available within walking distance are prepackaged, long-shelf-life, highly processed foods. Simple, complete foods such as fresh fruits or eggs are more expensive and harder to find for people living in poverty.

This shift has occurred for a variety of reasons. When President Richard M. Nixon was inaugurated as president in the late 1960s, he was warned that food scarcity and high food prices would be one of his main challenges.[29] Our nation was growing quickly, but agricultural productivity wasn't keeping up for a variety of demographic reasons. This risk of food scarcity is something that other countries face routinely, but it's hard for a modern-day American to imagine that it was part of our national policy discussion not so long ago.

In response, policies were developed to support increased production to make food abundant and affordable, and they prioritized grains. Even the Food and Drug Administration's nutritional guidelines switched from promoting a balanced diet from four food groups—dairy, meat, grains, and fruits and vegetables—to the pyramid that directly encourages significantly higher consumption of grains at the expense of meats, fruits, and vegetables. This switch was unfortunate, as higher carbohydrate consumption is correlated with some chronic diseases and obesity.[30] Indeed, a 2016 study by Seattle Children's Hospital found that removing grains, dairy, processed foods, and sugars from diets can bring pediatric patients with active inflammatory bowel diseases into clinical remission.[31] Direct and indirect subsidies for agricultural operations and higher-energy inputs to increase productivity on the farm made corn and its offshoots, such as high fructose corn syrup, much more abundant and affordable today. In parallel, we have a remarkable obesity epidemic underway in the United States and around the world.

The food abundance that modern energy enabled certainly solved many food scarcity challenges but also had some unintended consequences. Notably: portion sizes in the United States have been increasing over time. A standard 7-ounce soda at McDonalds in the 1950s (around the time they started a decades-long partnership with Coca-Cola) has grown to a standard size of 32 ounces or larger.[32] The fries and burgers have grown, too.[33] The Western diet includes more food calories of energy than ever. Though the rapidly increasing caloric content of our meals feels like a recent phenomenon, some elements of it have been around for a long time as part of a centuries-long trend. A study in 2010 of dozens of paintings depicting the Last Supper concluded that the amount of food served to Jesus and his disciples had grown significantly over the previous millennium.[34] So food inflation has been underway for a while, though fossil fuels let us put that in practice—not just art—in an accelerated way.

Not only is there a lot of energy embedded in the overproduction and subsequent overconsumption of food, but there is also more energy consumed because of our obesity.[35] We spend energy feeding the world's population of more than 400 million obese adults and then more energy moving ourselves around in planes and cars. Reducing our consumption of food to healthier levels would not only avoid a lot of unnecessary energy consumption, but also lead to us living healthier, longer lives.

In addition to reducing how much we consume, we should also reduce how much food we waste. Remarkably, in developed economies such as the United States and United Kingdom, about 25–40 percent of edible food is thrown away.[36] This level of food waste is stunning and an outgrowth of food policies that have promoted food abundance and affordability and a cultural shift away from thriftiness.

The amount of energy embedded in the edible food discarded in the United States is staggering, coming in at 2 quads or more per year.[37] That's enough energy to power entire countries like Switzerland or Sweden for a year. And that doesn't include the energy embedded in the inedible food we throw away. There are losses all up and down the food supply chain, including in farm fields, in processing centers, at grocery stores, and in our homes and restaurants.

Food waste is also related to American cultural phenomena such as all-you-can-eat buffets, where there is no price penalty for serving ourselves too much food. If we put too much food on the plate and then decide not to eat it, we don't suffer a financial consequence for that waste. In addition, most restaurants don't allow people to take home the extra food from buffets, exacerbating the situation further, though the rationale makes sense. If people could load their doggie bags with excess food, then they might be tempted to intentionally get an extra meal out of their tab.

Some people might reduce their food waste at restaurants by keeping their remaining food for leftovers, but my anecdotal observations of friends and family indicate there is also a common cultural distaste for leftovers. Some people just don't like day-old food. Becoming more comfortable with eating leftovers would go a long way toward reducing food waste.

Other cultural aspects of this problem include our focus on shopping for foods based on their visual appeal. Many perfectly edible foods are left to rot in grocery stores because they have bruises or look asymmetric. In response, some activists have launched an unlikely campaign in support of eating ugly fruit rather than throwing it out.[38]

In an era of abundance, we are more likely to become picky eaters. In an era of scarcity, we would eat whatever we were served. It's not an unusual refrain to hear from our grandparents about how we should clear our plates and that we cannot have seconds until we eat our firsts and that a starving child in some remote corner of sub-Saharan Africa would love to have the food we are not eating. That ethic of not wasting food took shape during the Great Depression, in an era when food scarcity was much more common, food was harder to get and pricier, and it was considered shameful to be wasteful. We have lost our cultural attachment to thriftiness with our food.

Reducing food waste can also have religious connotations. This anecdote, quoted from an essay by Vijaya Nagarajan, is based on observations made on the outskirts of Madurai, which is a large city in southern India:

> It was not unusual to see both men and women traveling frequently throughout the neighborhood streets and asking for food at each house. . . .
>
> In most households, meals were still prepared fresh; it was an enormous amount of work for the women of the house. . . .
>
> When a beggar would come to the door, then, it was the custom to feed the beggar, to give some rice, lentils, vegetables, sometimes cooked and sometimes raw. There was even a concerted effort to find a beggar, in order not to waste the leftover cooked food. From my upper flat balcony, I would hear voices of mothers calling their daughters to go down the street so that they could bring back a beggar to feed, so the food would not go to waste. Beggars were in demand, morally and physically, soaking up the excess food supplies. If you wasted food, it was

as if you were wasting god. The concept was offensive. The goddess Lakshmī is said to be embedded in every grain of rice.

A refrigerator had arrived some weeks before in my neighbor's house. It was a big moment, when the neighbors from several houses gathered together and shared in the festivities and joys of the first refrigerator coming into the neighborhood.

Some weeks later I observed the following scene. A beggar arrived at the door of the neighbor who had just bought the refrigerator. An onomatopoeic sound issued from the doorway: "Shoo, shoo, shoo, go away, go away, why are you people always coming around here?" I listened to the tone of voice.

A refrigerator-owning woman has, perhaps, a different relationship to leftovers than the non-refrigerator-owning woman. The freedom gained for the woman who no longer has to cook everyday can be juxtaposed against the loss of access to perishable food from the point of view of a wandering beggar. The very concept of waste as it relates to food undergoes a dramatic change with the coming of a refrigerator into a community. Generosity takes on a different character. Rituals of generosity, therefore, also change. Holding onto surplus food and storing it versus letting it move through the household and beyond the individual household becomes more prevalent.[39]

Some people have complained that fossil fuels—which gave us mechanized and industrialized agricultural systems—have removed us spiritually from the land, severing our connections with our food, which makes us less respectful of the animals who are slaughtered to feed us or the enormous inputs required to grow the food. They also reduced the manual labor required for food. Perhaps the same complaint can be made about electricity: By giving us refrigeration, it has extended the life of our food, reducing food waste and the labor of food preparation dramatically. These are good things. But we have also changed our relationship with food and those who need to eat. Perhaps the moral character of food has been lost, too.

## FOOD SECURITY

Remarkably, in this age of abundance, there are still at least 1 billion people who face food insecurity in some fashion. Even in the wealthy United States, there are millions who grapple with food insecurity. That often means that they don't know where their next meal will come from.

At the individual level, food security is a life-and-death consideration; at the national scale there is also the connection between food and national security, just as there are connections between energy and national security. Author Mark Kurlansky observed in his book *Salt: A World History* that because of salt's important role in preserving food, seventeenth-century British leaders spoke with the same kind of urgency about the country's national dependence on French sea salt as contemporary leaders speak about dependence on foreign oil.[40] Looking back, their concern about salt—which is abundant and cheap today—seems comical. Perhaps in the future we will say the same thing about today's fretting about oil.

The reason British leaders were so concerned about salt is because food security is a matter of national security. That remains true today, and food is a serious business for the US military. I experienced firsthand the intense effort and coordination required to feed hungry servicemembers in the cadet mess hall at the US Military Academy at West Point. The cadet mess hall is a glorious building built from large gray granite blocks in a style reminiscent of Gothic cathedrals. Six spokes radiate out from a central tower, each spoke comprising a sprawling, cavernous hall like the main room at Hogwarts School of Witchcraft and Wizardry in the Harry Potter movies—but bigger, more impressive, and without the floating candles. The scale by itself is breathtaking; it must be one of the largest dining rooms in the world.

The mess hall is not open to the general public or even the faculty at West Point—it's for the cadets. But honored guests who are invited to join the cadets for a meal are allowed to experience the phenomenon of lunch at West Point. That's what happened to me in April 2014 when I was on campus with my then 11-year-old son, David, to give a keynote lecture on energy and national security to the entire plebe class of nearly 1,100 cadets.

The entire student body eats lunch at the same time. To begin, all cadets stand at attention behind their chairs while announcements are made. When the order is given for lunch to start, the cadets quickly sit down at their tables and then a whirlwind of servers appear as if by magic from the backrooms of the mess hall and deliver trays of food to the more than four thousand cadets. Viewed from above, this must look like a finely orchestrated symphony of movements as food comes out to the tables and works its way down in family-style serving containers from the seniors to the plebes. And the whole thing, including announcements, has to take place in less than 25 minutes. My son and I were slackjawed as we watched food appear suddenly at the table, work its way to us, and get consumed hungrily by the eager students. Then, just like that, they were gone, hurrying off to their next destination. So much effort and thought goes into this meal—and this is just one meal at one location for one small fraction of the entire armed forces. Experiencing this lunchtime phenomenon served as a gripping reminder that food is critical to the military's success.

Food security is also tied to energy security, so energy disruptions or bad energy policies can disrupt the food supply, which can lead to unrest. For example, a food crisis triggered the tortilla riots in Mexico in 2007.[41] US biofuels policy caused corn prices to go higher, which raised the price of tortillas. While higher prices are good for farmers, they are bad for people who eat. That led to the riots. Similar problems occurred elsewhere. High energy prices (along with floods, droughts, and events that reduced food production) caused food prices to spike, exacerbating underlying tensions in the Middle East and ultimately helping to trigger the Arab Spring in 2010.[42] This wave of unrest ultimately toppled the government in Tunisia before spreading to Libya, Egypt, Yemen, Syria, and Bahrain.[43] In the United States we had our own security challenges tied to food. After Hurricane Katrina struck New Orleans in 2005, it knocked out the power. When the power went out, refrigerators and freezers quit working. Once the food spoiled, unrest in the city started to bubble over.

Our grandparents' and great-grandparents' attitudes about food security were also shaped by World Wars I and II. During both, food was considered a strategic asset and a national security issue. Wasting food was a way to support the enemy. But saving food, canning food, and not wasting a morsel could strengthen our troops and contribute to our victory. Consequently, a series of graphic posters and other propaganda art were used to educate and mobilize Americans to be more thoughtful about their food choices, chiding them to waste less.[44] Those posters encouraged shifts in diet—away from wheat and toward potatoes—so that wheat, which ships more easily without rotting, could be moved to the front lines. Nursery rhymes such as

> *Jack sprat could eat no fat,*
> *his wife could eat no lean,*
> *so betwixt them both, you see,*
> *they licked the platter clean*

were printed on posters, encouraging all of us "to lick the platter clean," leaving no wasted food behind. Some posters had a regional flair—noting the importance of Kansas food to the war effort, for example—but all accepted and propagated as an underlying premise the central role that food played in our security as a nation. Wasting food was the same as wasting energy, and both were needed on the front lines. Interestingly enough, in an era when Rosie the Riveter posters highlighted and celebrated how important women workers were for the war effort, the posters also put women in the central role as decision makers regarding food in the home: what would be eaten, how it would be prepared, and how it would be saved (for example, by canning).

## FOOD FOR ENERGY

The energy-for-food relationship goes the other way, too. We use a lot of energy for food, and we also use food for energy. Not just for our bodies, but also for our modern machines.

Ethanol from corn is a fine transportation fuel, but it has a variety of trade-offs. It is high octane, which means it can achieve higher compression ratios in the engine, which is a fancy way of saying it gives more power. Race car drivers love alcohol-based fuels because the extra power is a useful advantage. The Model T Ford ran on ethanol, making it one of the first dual-fuel vehicles available. But ethanol has lower energy density than gasoline, which means a gallon of it doesn't take a car as far.

It's domestically produced in the United States from corn, which takes $CO_2$ out of the atmosphere while it grows. However, it's much harder to grow biofuels—it requires more energy inputs and is much harder on the land in that it requires more space and is tied to runoff—than to extract fossil fuels from underground reservoirs. Biofuels do create wealth for farmers, which was an important political motivation for policies supporting their development, because US elected officials like to support rural midwestern communities, where the corn is often grown.

Corn-based ethanol involves moral conflicts as well. Because of congressional mandates that gasoline contain 10 percent ethanol, about half of all the corn that is grown annually in the United States is used as fuel for our vehicles, displacing about 10 percent of our annual gasoline demand.[45] That same amount of food could nourish tens of millions of people, but we use it to power our autos. We could have made our cars 10 percent more efficient using tighter fuel economy standards and off-the-shelf technologies to save the same amount of gasoline at lower cost while either avoiding the energy inputs and environmental impacts of all that corn growth or switching to strains of corn that would be suitable for feed or food. Since a significant fraction of our vehicles are gas-guzzling SUVs and trucks, it seems that we are using food—calories that hungry people want—for luxurious purposes rather than for necessities. This example of a heavy-handed—and perhaps wrong-headed—policy just goes to show that sometimes we stand in our own way, making the problems worse.

In Brazil, ethanol is also quite popular, but its growth and use avoids the moral conflicts and environmental conundrums typical of US production. Sugarcane is more highly productive, yielding more gallons of ethanol per acre annually than corn. Sugarcane plantations in Brazil yield seven harvests before replanting is required, in contrast with corn, which needs yearly replanting.[46] Sugarcane, once harvested, can be directly converted into alcohol by fermentation. Corn, a starch, first needs to be converted to a sugar before it can be fermented. That extra step is energy-intensive and costly. Sugar in Brazil is also rain-fed (instead of irrigated) and doesn't need the same energy inputs, such as fertilizers, to grow. When sugarcane is harvested, the bagasse—or fibrous part of the plant—is burned in a power plant to provide heat and power in place of fossil fuels, reducing the impact further. Finally, for reasons having more to do with the economic conditions of the country than a production preference, the sugar is harvested using manual labor rather than mechanized equipment. The work is backbreaking, and while its use avoids significant diesel consumption for tractors and other equipment, from a social justice perspective it would be better to use petroleum-based machines.

Sugarcane has been cultivated in Brazil for over 500 years.[47] That some locations have grown sugarcane year-in and year-out for centuries without significant soil erosion is a sign that this crop can be sustainably produced in Brazil.

After the sugarcane is produced, harvested, and converted into alcohol by burning the bagasse to provide heat for fermentation, it is made available to consumers. When Brazilian drivers arrive at a fueling station, they have the choice of buying ethanol or gasoline. While a small fraction of cars in the United States are flex-fuel capable—meaning they can operate on blends of alcohol or gasoline—nearly all new cars in Brazil have that feature. Consumers can simply choose whichever fuel they prefer—usually the cheaper one. Doing so means gasoline prices keep ethanol prices in check and vice versa. It also means renewable ethanol is now responsible for half of the liquid fuels consumption in Brazil, a remarkable outcome of their biofuels policy.

Not only is the sugarcane ethanol produced in a cleaner and less impactful way than corn-based ethanol, but it also avoids the moral conflict about using food for fuel. Sugar is an important source of calories, but in Brazil and elsewhere, sugar is primarily an additive for flavor rather than a main course. The Brazilian experience shows that producing biofuels on a large scale is difficult, but it can be done. At the same time, Brazil has had more than four decades of steady research and

development, investment, and policy making that helped in-
centivize that very outcome.

There are also ways to replace petroleum-based diesel with
biodiesel from foodstuffs. In fact, the first diesel engine was
demonstrated by Rudolf Diesel in 1893 using peanut oil as
the fuel. That means the diesel engine came to life on the energy of food. More than a century later, in late 2017, London's
iconic red double-decker buses began using oil extracted from
ground coffee waste for its fleet, in blends that are 80 percent
petroleum-based diesel and 20 percent coffee-based biodiesel.[48]
This approach puts a waste stream to use while also reducing the
carbon footprint of the buses and provoking snarky comments
about whether passengers would arrive to work in the morning
more alert than normal because of their caffeinated journey.

Today, biodiesel is made from feedstock other than peanut
oil. In the United States biodiesel is made from soy, which is
one of our staple crops (along with wheat and corn) and is a
common source of protein nationwide. Europe and other major
consumers use biodiesel made from palm oil.

But palm oil, while abundant, comes with its own significant trade-offs. It is often produced on plantations in Malaysia
or Indonesia. To create the plantations, which are very profit-
able, the native jungles are first burned down, emitting a lot of
air pollution and carbon dioxide in the process, undoing many
of the environmental gains that biofuels are intended to offer. In addition, palm oil is an important source of calories for
many people, which exacerbates the food-versus-fuel debate.
The deforestation in Southeast Asia is also reminiscent of deforestation in the United States in the 1800s, when wood was
needed as a building material and source of fuel for heating
homes, cooking, and industry. That rampant deforestation was
stopped when fossil fuels—namely coal—came along. But using
biofuels to replace a fossil fuel (in this case, petroleum) seems

to take us in the opposite direction, reintroducing the risk of deforestation.

Deforestation can also occur as an indirect result of biofuels policy. Europe has prioritized biodiesel from palm oil as a fuel for its more common diesel engines, and the United States has prioritized ethanol from corn for its gasoline engines. Because of the new demand for corn ethanol, US farmers have increased production of corn and reduced production of soy. After soy production dropped, global soy prices increased. Brazilian farmers, chasing the higher soy prices, bought land from ranchers that had been used for grazing and planted soybeans. Those ranchers then burned down more Brazilian jungle to gain access to land for grazing.

But biodiesel does not have to be made directly from food. It can be made from food waste. The grease used in restaurants can be collected and, with very little treatment—not much more than filtering out the solid bits—burned in a conventional diesel engine. In the neighborhood in Southern California where I lived for many years, a resident of the neighborhood had outfitted his Volkswagen Jetta to run on waste cooking oil from a local Mexican restaurant, which was a tidy environmental solution to a waste product that also displaced his consumption of petroleum. It also had the side benefit that every time he drove by, his exhaust fumes smelled like tortilla chips.

Turning food waste into energy is a good idea beyond just the oil from Mexican restaurants. There are a few ways to do this. Food waste can be composted, which converts it into a valuable soil amendment. Composting carefully is a sensible, natural way to reuse those scraps, but also poses its own risks to the environment. People who compost poorly might put a lot of methane into the atmosphere, which is both a blessing and a curse. While it's a valuable feature that organic materials (such as most food wastes) can be converted into methane when decomposing anaerobically (i.e., in a sealed container with very little or no oxygen present), methane is a very potent greenhouse gas. But methane is also a very useful fuel and is the primary constituent of natural gas. That means food wastes can be turned into a biogas that can replace natural gas in power plants, industry, and for home heating or cooking applications. In a country like the United States, which wastes so much food already, this idea could scale up well. We already do it accidentally and slowly when food scraps in the garbage ultimately make it to a landfill. As the layers of trash are covered with dirt, the supply of oxygen is cut off to the food waste. Consequently, as the food in the landfill breaks down, it is converted into methane. This methane can be harvested intentionally via vents drilled into the trash mounds to collect what is commonly known as landfill gas. Or we can speed up the process and control it more tightly by directly converting food waste into biogas using anaerobic digesters—large, shiny metal tanks with a suitable dose of microbes—skipping the landfill entirely and giving a steady stream of renewable natural gas.

I can imagine a trash bin, a recycling bin, and a food waste bin on the street in front of each house in our neighborhoods, and trucks collecting our food waste to create a concentrated stream rich in organic materials that could be converted into biogas at a dedicated facility, increasing efficiency and bringing down costs. Or maybe we switch to a food delivery system that brings food to people's homes but also takes food waste back to the food warehouse. That means the same warehouse can sort food for delivery and then sort food waste for conversion into biogas that would either create another revenue stream or reduce energy costs on-site. The warehouse could also convert whatever spoilage is available on-site into energy, recovering product for a useful end. Our research at the University of Texas at Austin indicates that the extra energy expenditures for delivery are offset by even greater savings from avoiding food waste. But the packaging in which the food is delivered and picked up increases energy consumption, offsetting those benefits. That means packaging matters: reusing containers, like the old glass milk bottles delivered to my house when I was a kid, would save energy.

In Austin, we can put our food scraps down the drain. The city of Austin's water and wastewater utility installed industrialscale anaerobic digesters. When we put our food down the drain through garbage disposals, it flows down the sewers with our other waste from toilets and showers to be collected and treated. Those large digesters convert the organic waste en masse into soil amendments called 'Dillo Dirt (in homage to our local creature, the armadillo) and significant volumes of biogas. Thousands of tons of biosolids are turned into 'Dillo Dirt each year and sold through local vendors for landscaping and other applications.[49] This same concept can also be implemented on the farm: collecting crops that aren't suitable for market and otherwise would rot in the fields, and breaking them down in the digesters to produce biogas. An even better opportunity is to convert the 100 million tons of manure produced every year in the US by cattle, pigs, and poultry into methane that could be used to offset dirtier fuels, reducing emissions in the process.

Managing that manure is a major hazard for farmers. In prior centuries, farms with a mixture of crops and livestock would spread the manure over the nearby fields as fertilizer. But fossil fuels have enabled a densification of food production, which separates agriculture into two problems: a fertilization problem for crops and a waste-handling problem from livestock.[50] For animals, these are called concentrated animal feeding operations (CAFOs). Now instead of having a mixture of pigs, poultry, and cattle along with a range of crops, large farms often grow only one crop or raise one type of livestock. And because of productivity gains and efficiencies that came along with fossil fuel use, these facilities are much larger. Large feedlots can have 100,000 head of cattle, and large poultry operations can have millions of chickens that each produce an egg daily. Such scale would not have been possible in the pre–fossil fuel era.

This concentration of industrialized food production activity is a modern marvel. But it also creates environmental hotspots from concentrated waste streams. The volume of manure that is produced far exceeds what the local fields can accommodate, creating a major challenge for farmers to manage. It is too expensive to simply truck the manure away to far-flung locations, so it must be managed on-site.

Manure is usually stored in man-made lagoons lined with a thick tarp of plastic or some other barrier to prevent the waste from seeping into the ground and contaminating the soil or groundwater. Those lagoons, storing the waste from thousands of animals, are a major source of odor and a costly liability for the livestock operations.

When the manure mixes with urine, it produces significant quantities of ammonia ($NH_3$). When ammonia mixes in the air with other pollutants typical of power plants, smokestacks, or vehicle tailpipes such as $NO_x$ (nitrogen oxides) or $SO_x$ (sulfur oxides), it forms ammonium nitrates and ammonium sulfates, which are types of fine particulate matter that can get into the lungs of workers and animals. That makes the manure piles and lagoons an environmental risk in several ways. As part of my work making environmental sensors right after graduate school, I invented a sensor that could measure trace pollutants such as ammonia coming from manure in agricultural operations.[51] In other words, I invented a bullshit detector.[52]

But that manure is rich in organic materials. Just like food waste, the manure can be a source of energy. As before, manure can be converted into biogas through anaerobic digesters.

These large metal objects use microbes and heat within the manure to convert it into biogas and a solid leftover known as digestate that can be used as fertilizer. Because the lagoon is anaerobic underneath the surface, it is also not unusual for the manure to break down within the pond, forming massive methane bubbles. Sometimes those bubbles pop, sending manure flying, which must be quite a sight. Some feedlots also pile the manure into miniature mountains, which must be managed so that they don't catch fire.

If farmers create and capture that biogas intentionally, they can sell the gas as a renewable energy resource or convert it into electricity on-site with gas turbines, small generators, or fuel cells. Doing so lets them convert an expensive environmental liability (the manure) into a new revenue stream (either gas or electricity). By our estimations, there is enough manure in the United States to meet 2 percent of our electricity needs.[53] At the same time, the methane that would have been vented into the atmosphere, where it would be a very powerful greenhouse gas, would be instead burned and converted into carbon dioxide ($CO_2$), which absorbs much less solar radiation. Using manure, an important by-product of the food system, as a source of energy solves multiple problems simultaneously.

In addition to the solid waste produced as excrement from the animals, cattle produce a lot of methane from their normal feeding and digestion processes. It has often been suggested that feedlots should capture this rudely produced methane as a renewable source of energy. It could be called "Fuel Alternative Research Technology," or FART for short. (Note: it is actually cow burps from digesting their food that produce so much of the gas, so the joke falls short for knowledgeable audiences.)

Another challenge with agricultural waste is crop residue, whose disposal can create significant environmental impacts. In India, nearly 100 to 140 million tonnes of crop residue are burned annually, which creates breathtaking smog for New Delhi and other major cities. A colleague of mine at the University of Texas at Austin, Dr. Vaibhav Bahadur, has proposed that these crops should instead be gasified into methane to power up an off-the-grid refrigeration system to condense water out of the atmosphere.[54] Doing so would help farmers gain water they need for irrigation, while reducing their costs and reducing pollution in cities. One tonne of biomass can generate 800 to 1,200 liters of potable water. Converting that biomass into energy for water harvesting would meet 7–10 percent of potable water requirements in various Indian states, which is particularly relevant to agrarian locations with abundant biomass but a scarcity of potable water.

The energy sector's waste streams could also be part of the food system. Power plants produce significant volumes of $CO_2$. At the same time, society generates a lot of nutrient-rich wastewater. Nutrient-rich water and $CO_2$ are critical inputs to photosynthesis. So it is possible we could combine these two waste streams, using dirty water from our sewer system along with $CO_2$ from our smokestacks to grow algae. The algae simultaneously consumes the $CO_2$ and cleans the water while creating biomatter that can be used as a source of protein for nourishment and a source of lipids for biodiesel. By coupling the waste streams, we can grow algae as a source of feed, food, fuel, and freshwater.[55]

It has even been proposed that we can capture the $CO_2$ in power plant flue gases, using a method that converts it into baking soda that can be used to make cookies. The implication is that we could eat our way out of our energy problem. However, each tonne of $CO_2$ makes 2 tonnes of baking soda. The world produces about 30 billion tonnes of $CO_2$ each year. If we converted all of that $CO_2$ into baking soda, it would flood the world's markets, causing baking soda prices to collapse, and perhaps triggering an obesity epidemic if we ate all those cookies.

In the end, the story of using food as a form of energy is a helpful reminder that no energy solution is one-size-fits-all. Each option is beset by trade-offs, and there is no magic bullet.

## THE FUTURE OF ENERGY AND FOOD

Given all the energy inputs and environmental impacts of the food system, it is worth considering ways to reduce its energy inputs while also maintaining high levels of productivity to meet global food needs. Because there are still more than a billion people facing some level of food insecurity, investing in the productivity of the food system is still important. Investing in energy and energy-enabled technologies is part of that suite of solutions.

Improving production on farms can be accomplished with new technologies like laser-leveled fields to reduce runoff of water, soil, and nutrients, and advanced irrigation systems that reduce how much water is needed to improve crop growth. Rather than simply flooding the fields, which saturates the soil in places and causes leakage of fertilizers or other chemicals to the water table, or spraying water into the air only to watch it evaporate on the way to the soil, there are drip irrigation and subsoil irrigation systems that are much more efficient. In other parts of the world, where food is still harvested by hand, investing in new equipment and fertilizers could offer a big boost.

Tractors and other mechanized field equipment represent an opportunity to reduce the greenhouse gas emissions from and energy requirements of farming. Because tractors drive a limited range (in the dozens of miles per day), have space for large batteries, and often return to a resting place each night (e.g., a barn), they are good candidates for electrification. Electric motors are more efficient than diesel engines, so that switch, even when accounting for the losses at the power plant that generates the electricity, would save energy consumption overall. Electric motors would also be quieter and produce fewer emissions. Those tractors could also be programmed to run without drivers according to a prescribed field pattern using GPS coordinates. That approach would reduce labor and time.

Some cattle ranchers are also working on raising low-carbon beef, which involves breeding, using different feeds, altering feeding cycles, and managing manure with the intent to reduce inputs and reduce emissions. For example, researchers at the University of California at Davis have found in preliminary trials that using seaweed as a dietary additive can reduce enteric methane (that is, the methane from cow burps) by 50 percent or more.[56]

Farmers might eventually be paid to put carbon dioxide back into the soil. As soil is turned over, carbon dioxide is released into the atmosphere. But with a few different soil management programs $CO_2$ can be withdrawn from the atmosphere and sequestered in the soil.[57] Programs could be created to encourage these options, and we should pay farmers for their services when they help solve the climate crisis.

Another advance looming for the food system is the growing popularity of food delivery services. Companies like Blue Apron and others deliver prepared meal kits to homes, and grocery stores have expanded their delivery offerings. At first blush it seems much more energy will be consumed with this approach, because it means delivering food via trucks with low fuel economy that potentially operate on diesel engines. But it might also mean fewer car trips to the grocery store by shoppers. If we restrict our store shopping to foods with long shelf lives such as flour, sugar, and cereals, and nonfood items like paper towels or other cleaning supplies, then going to the store on a monthly basis could be a possibility. And that truck delivering food to our house might be delivering to our neighbors' houses, too, saving many of us trips to the grocery store. The truck might also run on electricity or natural gas, which are quieter and cleaner for our residential neighborhoods. Drones might drop the food right on our doorsteps. Importantly, meal kits delivered to our homes might reduce food waste dramatically because the meal portions would be sized for what we select to eat, though the packaging would still need to be recycled or reused to offer an energy benefit. Rather than over-buying our perishable foods at the store to minimize the number of trips we need to make and then letting that food spoil at home when we don't eat it, meal delivery lets us fine-tune our ordering to our desired eating schedules rather than to our desired shopping schedules. Because there is so much energy embedded in food, reducing the food waste would potentially save more energy than is required for the delivery service.

Other technologies could be implemented along the cold chain to maintain temperatures that prevent spoilage. That is, by spending more energy on refrigeration in warehouses, trucks, and train cars, the energy loss from food waste might be avoided. Maintaining better control of the temperature conditions of food would have a lot of benefits. Food spoilage becomes a public health issue when produce or meat becomes contaminated by *E. coli* or other pathogens. As a consequence, food purveyors and manufacturers use labels with "sell by" dates to reduce the chances that consumers unwittingly buy food past its prime. However, these labels are very clumsy indicators based on predictions rather than the actual status of the food in the package, and consumers are easily confused by them.[58]

According to a joint study by the Natural Resources Defense Council and the Harvard Food Law and Policy Clinic, up to 91 percent of consumers prematurely discard food, and misinterpretation of date labels is responsible for 20 percent of wasted food in British homes.[59] This food waste is avoidable with better labels. Of fifty possible solutions to reducing food waste that were analyzed, printing standardized date labels on food is the single most cost-effective approach because it would reduce consumer confusion. According to ReFED, a nonprofit dedicated to reducing food waste in the United States, better food labels alone could avoid 400,000 tons of food waste annually, saving nearly $2 billion.[60]

Many foods are safe to consume far beyond the sell-by date—especially if they are vacuum-wrapped and frozen—but it is also possible for foods to spoil before their printed expiration date, for example, if they are accidentally exposed to higher temperatures for many hours during transit because the cold chain had a mechanical failure. Instead of printing dates on food labels, it would be better to incorporate miniature sensors into packaging that track the temperatures the food is exposed to. These sensors could use temperature-sensitive inks so that if the food is ever exposed to too high a temperature for too long, the label changes color to let you know the food is unsafe to eat. But if it's maintained at the proper temperature, the label stays the same color. If the food is frozen, then the label maintains its safety indication even longer, but it might also tell you the food was frozen, in case it's important to a consumer to eat food that has never been frozen.

It would also be helpful to have other technologies beyond refrigeration that help avoid spoilage. Many people are familiar with the pithy maxim that "one bad apple spoils the whole barrel." This concept has been used as a teaching proverb related to sinners and bad behavior, but it turns out this concept has a scientific basis. When apples rot, they produce the gas ethylene, which causes apples and other fruits to ripen. As the first apple goes bad, it emits ethylene, which triggers ripening in neighboring apples such that they also produce ethylene. The riper a fruit is, the more ethylene it produces, compounding the effect.

So, indeed, one bad apple can spoil the others. One way to mitigate this risk is to tightly manage the quality control so that no bad apples are mixed with the others. Another way is to install filtration systems that monitor and remove ethylene as it is produced so that its ability to trigger ripening is reduced. Using carbon filters (made from coal) and circulation fans (operating on electricity) to clean the air is just another example of how energy can be used to reduce food waste, thus saving energy.

There are also some passive technologies that help. Vestergaard, a Swiss company that makes the LifeStraw for water treatment, also makes the ZeroFly food storage bags for farmers to use to store their grains. In the United States, most food waste occurs at the consumption end of the supply chain because we serve ourselves portions that are too large or buy too much food. But in the developing world, most of the food waste occurs closer to the farms themselves. Standard pests like weevils, molds, and fungi can wipe out a farmer's economic livelihood. These farmers often lack the resources for the type of high-end, shared storage silos one might find in Minnesota. Instead, they store their grains in simple sacks in their homes, and if the food spoils before they can sell it, then they are out of luck. ZeroFly bags can be hermetically sealed, and the bags are coated with insecticides. The insecticides keep the pests from burrowing through the sack. And because of the hermetic seal, the bag's interior is oxygen-starved. Any critters that need oxygen to survive will die. This simple technology—it looks and feels just like the other bags they have used for decades—can reduce food waste dramatically while giving those farmers a chance to achieve better economic returns.

In the end, energy and food are heavily intertwined. They are even synonymous because the food system requires energy, food is a form of energy for our bodies, and the modern energy system is starting to rely on food as one of its sources of supply.

Changing our energy system gave us a food revolution, which is the model for going the other way around. Changing our relationship with food could have profound effects on the energy system. Water is also critical to the food and energy systems, so the three combined form a special foundation for society: any impacts on the integrated food-energy-water system will spread out to everything else, including transportation, which is the next priority for a modern society. With the right combination of policies, markets, and technologies, we should be able to move the needle to increase food supply while reducing energy intensity, with widespread benefit to the world.

Chapter 3

# TRANSPORTATION

The desire to explore is a defining aspect of humanity. That makes transportation a very human undertaking. And it depends on different forms of energy and energy conversion devices, such as engines or motors, to happen.

In prior centuries, transportation was achieved by manpower (rowing boats or walking), animal power (pulling horse-drawn carriages and dogsleds), and wind power (sailing ships that traversed the globe). In the thirteenth century, Friar Roger Bacon, a monk who was credited with being one of the developers of the scientific method, made some stunning predictions: "Instruments may be made by which the largest ships, with only one man guiding them, will be carried with greater velocity than if they were full of sailors. Chariots may be constructed that will move with incredible rapidity without the help of animals. Instruments of flying may be formed."[1] Even several centuries ago, when transportation was a purely physical affair, we dreamed of machines that would take us farther and faster without burdening our muscles.

These fantasies from the thirteenth century became reality because of the magic of energy and the invention of the gasoline and diesel engines in the late nineteenth century. The basics of those engine designs are still the primary drivers of transportation today.

In Bacon's era and the millennia before and the centuries that followed, the ability to travel beyond where our feet could take us was mostly the domain of the rich who could command sufficient manpower and the fleets of ships or stables of horses necessary for longer expeditions. As a result of some of the grandest technical achievements in human history, travel is no longer the exclusive enclave of the very rich, powerful, and adventurous: energy has liberated middle-class and poorer people from the confines of their hometowns. Coal-fired trains, gasoline-powered cars, and jet-fueled airplanes connect the world's nearly 8 billion people. The miracles of energy have ensured that there is always a way to get almost anywhere in the world. A select few humans have even traveled to the moon, thanks to rocket fuel. But transportation moves more than just people around. We depend on it to get food, medicine, and goods.

Transportation has shaped our societies: where we can feasibly travel in the course of our day-to-day activities orders a great deal of our lives. And as our vehicles change, our concept of place changes along with them. Transportation has given shape to countries: canals did it in the early 1800s, rail did it in the late 1800s through the first half of the twentieth century, and highways did it in the late 1900s. Autonomous vehicles will change the layout and rhythm of our lives yet again.

Entwined with this story of change are the fuels and forms of energy we used to enable the engines and motors to propel us forward.

## WATER, WIND, AND MUSCLE POWER

Before modern forms of transportation were available, humans relied on water, wind, and muscle power for movement. Where waterways weren't available and before the steam engine made trains a possibility, muscle power was a key mode of transportation. Sometimes it was as simple as a person walking from one place to another or using the muscle power of horses or other draft animals. Muscle was also used for waterborne transportation, including the gondolas in Venice, Italy, powered by a single oar; punts on the Cam River in Cambridge, England, where those shallow boats are pushed along with a pole; or the dugout canoes or reed boats that have been around for thousands of years, propelled by paddles. Improved designs of paddles and oars later allowed teams of people—oftentimes slaves or prisoners—to row larger ships moving more tonnage. Ultimately, sailing ships were developed that harnessed the wind, liberating seamen from the need to do the hard work of propelling their boats and extending the range of seafaring people.

Because water is so efficient at moving goods, especially heavy goods, humans have tried for centuries to use waterways to transport items as far inland as possible. Canals were built as a way to simplify and enhance inland transportation. They could be designed with a series of locks to overcome altitude differences and prevent rapids, which would destroy boats. Canals could also be designed with pathways alongside so that horses or other forms of muscle power could tow the boats.

Because canals enhanced trade by reducing the cost of transportation, much of the economic growth of the United States during the 1800s was tied to their development. Of particular importance was the Erie Canal, completed in 1824, which connected the Great Lakes to the Atlantic economy through New

York. The canal shaved a significant amount of time off the journey, cutting costs to move goods from $125 a ton to $6 a ton in the process.[2]

Because of their importance, Abraham Lincoln's presidential candidacy highlighted investment in canals as part of his suite of campaign pledges.[3] Transportation projects like canals were attractive in the 1800s—just as transportation projects are still attractive today—because they create jobs during their construction and facilitate economic activity for decades afterward. Transportation and the economy were transformed again when cargo moved to rail in the late 1800s and then to trucks in the 1900s.

Canals and waterborne transportation followed by trains in the United States sprang up at a time when the nation was rapidly expanding and sharpening its identity as an industrial powerhouse. As William Cronon described it, these two transportation modes were transformative for the American landscape, especially for the West: "the central story of the nineteenth-century West is that of an expanding metropolitan economy creating ever more elaborate and intimate linkages between city and country."[4]

The rise of the modern city would not have been possible without transportation to bring goods in and out. Chicago was the first manifestation of this form of modern, energy-enabled city. Its rise was facilitated by a key geographic fact: it was situated between two major waterways—the Mississippi River and the Great Lakes (and the St. Lawrence River)—that were navigable. Close proximity to abundant natural resources—soil, prairies, and forests—and the systems to move them and their derivative goods made Chicago in the late 1800s "the greatest metropolis in the continent's interior, with all the Great West in some measure a part of its hinterland or empire." Because of these transportation linkages, at its peak the Chicagoan empire looked as expansive—and full of potential—as the Roman Empire had centuries before.

## THE STEAM ENGINE

Modern energy changed our transportation capabilities because it fueled the steam engine, which liberated transportation from wind, currents, and muscle power. The steam engine rose hand in hand with the industrial revolution because of its breakthrough ability to turn heat into motion. It is very easy to make heat by burning a whole variety of fuels, such as coal, charcoal, peat, wood, straw, and cow dung. When heat was used to make hot, high-pressure steam in a boiler, the steam could be harnessed to drive the motion of industrial equipment. At first it was used for devices like water pumps, but ultimately it was applied to other factory machines and transportation vehicles such as steam-driven trains and steamships.

Energy was the motivator for the steam engine, an enabler of it, and a beneficiary of it. The first practical steam engines were developed in 1712 by an Englishman, Thomas Newcomen, to pump water out of mines. As mines were dug deeper into the earth, water pooling at the bottom was a nuisance that made mining more difficult and expensive. But his design solved this problem by using steam to drive a piston that pushed a beam up and down that operated the pump. As metallurgy improved and cast iron or steel components were developed, the steam engines could be made larger and stronger.

But it was James Watt, a Scotsman, who improved the steam engine such that it could be used for more applications. His contributions to engineering and the science of thermodynamics are reflected in the unit of power, the watt (W), that is named for him. He came up with a design that improved the engine's power and efficiency dramatically. While the back-and-forth motion of early steam engines was convenient for pumping, Watt also configured the engine to produce rotary motion. Rotary movement is particularly useful for spindles, so those steam engines were adopted by the textiles industry, helping launch the industrial revolution. They also enabled the rise of steam-driven trains and ships, because rotary motion could power the train's wheels or steamships' paddlewheels.

One way to think about a steam engine is to envision it as an external combustion engine. The combustion of the fuel takes place outside the chamber (the boiler) that circulates water. As the water circulates and heats up in the boiler, it becomes steam, which can be used to push pistons or spin fan blades. Engineers figured out that the higher the temperature and pressure of the steam, the more force they could extract from it. This realization created a race to design ever-more-rugged boilers and materials that could withstand higher pressures and temperatures without melting or exploding, as burst boilers could kill and maim those nearby. A steam engine can be distinguished from a modern diesel engine because it has two plumes: a black one from the burning coal and a white one from vented steam that is used to regulate the internal pressure.

The steam engine made it easier to move more goods faster than animal transportation or sailing boats when it was incorporated into steamships, like those along the Mississippi River featured in Mark Twain's *Huckleberry Finn*, or into steam-powered trains, which transformed the landscape because they could reach landlocked destinations far removed from waterways suitable for navigation.

Chicago shows up again as a city where trains fundamentally shifted its relationship with the surrounding regions. Cronon wryly observed that "the railroads made Chicago, a city located in one of the nation's most treeless landscapes, the greatest lumber center in the world." With trains, finished lumber could be readily shipped to customers, so Chicago became a prominent location for lumberyards and sawmills. The ease and speed of trains also revolutionized the way farmers moved goods to market, as Cronon summarizes here:

> Once farmers had access to a railroad, most no longer thought it worth their while to spend a week or more driving a team of horses over bad roads to sell their crops in Chicago. More than twice as much wheat came to Chicago in 1852 via the Galena and Chicago Union [railroads] than came in farmers' wagons, the latter having fallen by half in just the previous year. In 1860, Chicago received almost a hundred times more wheat by rail than by wagon; ten years later no one even bothered to keep statistics on the latter.[5]

Trains changed the landscape because they broadened the reach of the American empire. Cities would pop up seemingly out of nowhere because the train arrived, opening up its connections to other locations. Dallas, Texas, is one of the nation's largest cities. But its rise was almost accidental. In the 1870s, a train line from Houston was under construction by the Houston and Texas Central (H&TC) Railroad to northern Texas and was routed to go through Corsicana before continuing on to Dallas. Dallas city leaders used legislative tricks to force an east-west line, the Texas and Pacific (T&P), through Dallas instead of Corsicana.[6] That made Dallas the first rail crossroads in the state, establishing it as a major shipping center. As a consequence, Dallas became a major metropolis, and Corsicana is a small town very few people have heard of outside Texas.

The steam engine was also used for personal transportation. Early automobiles used steam to drive a piston system connected with gears to the wheels to propel it forward. The most famous example was the Stanley Steamer, invented by the Stanley

Brothers, Francis and Freelan. The car had a water tank that would need to be refilled as the steam was slowly vented during operation. A sales brochure for the Model 735 seven-passenger car noted that it had a 24-gallon water tank (enabling a range of 150 to 250 miles before refilling), and a 20-gallon kerosene tank to heat the water in a boiler to make steam.[7] The car set all sorts of speed records, achieving 130 miles per hour in one demonstration, and was very popular. The Stanleys made so much money from their car manufacturing business that Freelan built an iconic hotel in Estes Park, Colorado, that still stands and was the setting for the horror movie *The Shining*, based on Stephen King's novel. An operational Stanley Steamer still sits in the hotel lobby today, and a steam-driven bus helps ferry visitors.

## THE INTERNAL COMBUSTION ENGINE

All societies motorize as they become affluent. While horses, then rail, and even the original steam-powered automobiles freed people to travel greater distances, the availability of automobiles powered by the internal combustion engine accelerated the movement toward personal transportation.

The internal combustion engine was another critical advance in converting heat—in this case from burning liquid fuels such as alcohol, gasoline, or diesel—into motion. The bulky steam engine powered by coal was a good fit for heavy vehicles like ships and trains, but other than the curiosity of the Stanley Steamer it was an awkward fit for moving vehicles at the scale of an individual or a family. Gasoline-powered engines were much lighter and more powerful than their steam-powered cousins and therefore a good fit for personal transportation. They started a revolution. The impact of individualized, mechanized transportation is hard to fathom. In total, it created some of the world's largest industries: automotive manufacturing (with steel made from coal and factories operating on electricity), oil production (for gasoline and diesel fuels), and road construction (using asphalt from petroleum and cement made with coal).

Think back to the family in the 1960s sitcom the *Beverly Hillbillies*, who found oil on their land and become wealthy overnight. That energy enabled not only the family's social mobility (letting them move up the ladder of affluence) but also their physical mobility. They used energy (in this case oil) to get rich, and then used oil to move across the country. People who have access to transportation have access to freedom.

Steam-powered cars needed two tanks: one for fuel like kerosene, and one for water that could be turned into steam. In this design the fuel was used for heat, and water (in the form of steam) did the work to drive the pistons. The fuel gave the heat, and the steam gave the motion. The brilliance of the internal combustion engine is that the fuel was the source of the heat *and* the motion. Instead of the fuel being burned outside the piston-cylinder that drove the forward motion, as in a steam engine (the steam was on the inside; the fuel was on the outside), the fuel in an internal combustion engine was burned inside the piston-cylinder system. As the fuel burned, the hot gases it produced had the force to move the piston. This design was simpler, more powerful, and lighter.

English and Scottish inventors led the world on the steam engines of the 1700s, and German inventors led the world on internal combustion engines in the late 1800s. Nikolaus Otto invented the four-cycle engine operating on liquid gasoline in 1876. Gottlieb Daimler of Daimler-Benz fame was an early adopter of variants of the Otto cycle, helping it to become the most common car engine in the world today. Modern versions have many performance enhancements, but the basic configuration is the same. The advances include in-cylinder compression and a spark for ignition that yielded more power in a smaller space.

Rudolf Diesel was another German inventor with a knack for improving engines. Rather than using a spark to ignite the fuel-air mixtures, Diesel conceived a design in the 1890s that used compression from the piston-cylinder system to cause auto-ignition. By using greater compression to raise the temperature and pressure of the gases, the fuel-air mixture would heat enough to burn automatically, even without a spark, which made the engine more efficient. But because the diesel engine operates with higher pressures and more compression, it requires stronger and larger engine blocks to avoid failure. That made diesel engines heavier and therefore more appropriate for large machinery such as trains, ships, construction equipment, or large trucks that needed a lot of power and for which the size did not matter. For smaller vehicles, like personal cars, or for planes where weight matters, the lighter, smaller gasoline engines using spark ignition were a better fit.

Petroleum was already an important resource because it provided kerosene for lighting, plus tar, wax, and asphalt. It became even more important because of the gasoline and diesel contained in each barrel. About half of a typical American barrel of crude oil comes out of refineries as gasoline, and about a quarter of the barrel is converted to diesel. The rise of the internal combustion engine fundamentally changed the importance of oil as a commodity. John D. Rockefeller was already the world's richest man from selling kerosene, and the advent of the internal combustion engine made him even richer.

Petroleum provided more than just the fuel: oil was also used as a lubricant for high-precision engine operation, and the rise of advanced metals for engine blocks (made with machine tools that also needed energy for their creation and lubrication for their operation) allowed more powerful engines to be created.

That is one of the ongoing trends of energy-enabled innovation: it usually makes things better, safer, cleaner, and more efficient with time.

The internal combustion engine can also run on biofuels such as corn-based ethanol or biodiesel made from soy. Biofuels have some advantages, but petroleum-based fuels have higher energy density and are cheaper and easier to produce on a large scale. As a consequence, more than 95 percent of transportation fuels today come from crude oil; despite various efforts to replace petroleum with biofuels, doing so remains a vexing challenge.

Just as early coal deliveries for home heating were made by horse-drawn carriages, early deliveries of fuels for cars were made by horse because they were considered more reliable than the automobiles themselves and a national trucking or pipeline infrastructure had not been built yet.[8] Over time, fuel distribution systems and the fuels themselves were improved. Higher octane in the fuels enabled higher compression ratios in the engine, thus leading to higher performance. The combination of improved fuels and engines meant that cars achieved longer range and better performance as time went on.

Though German designers gave us the first great engines, the United States was the place where cars became a way of life. We had the open spaces to enjoy them and the access to cheap domestic oil that made them affordable to make and operate. Mass production of the Model T Ford in the early 1900s helped brings cars to the masses, but World War I and then the Great Depression slowed their adoption. It wasn't until after World War II—as the economy rebounded and consumers had the money to spend—that car ownership took off and cars changed the shape of modern society. The post–World War II economic boom, America's success in the war, and the personal freedom associated with driving all seemed to go hand in hand to form the nostalgic concept of the 1950s American Dream. In 1955,

General Motors, Chrysler, US Steel, Standard Oil of New Jersey (the forebear of Exxon), Amoco, Goodyear, and Firestone were among the top ten employers in the United States.[9] The others were General Electric (which made technologies for the power sector, among others), CBS, and AT&T. That means most of the largest employers in the United States at the time were companies that made cars or the steel that was used to make the cars, the oil to operate the cars, or the tires the cars needed to roll. In many ways, the 1950s were the peak era of American auto manufacturing dominance. Historian David Halberstam claimed that General Motors was the most important company in America because of these converging factors.

Cars also enabled the rise of another icon of the 1950s: the suburbs.[10] In poorer eras, urban populations used streetcars, trolleys, or other means of transportation. After cars became affordable, a person no longer had to live within walking distance of work or the mass transit systems that would take them there. That gave families more flexibility to choose where they wanted to live for other reasons—proximity to relatives, access to parks, affordability, and so forth—than proximity to work.

Just as cars gave people the freedom to choose where they wanted to live, it also gave them the freedom to roam the country. Tourism and long-haul travel had been prohibitively expensive, but personal automobiles lowered the cost and time required for travel. Thus the great American road trip was born. But the quality of the roads advanced less quickly than the cars that used them, and road quality varied dramatically from state to state. State highways and federal highways were built to solve that problem.

The most famous highway was Route 66, a paved pathway from Chicago, Illinois, to Los Angeles, California, passing 2,000 miles through cities such as St. Louis, Tulsa, Oklahoma City, Amarillo, and Albuquerque along the way. My father grew up just a few miles from Route 66 in Monett, a small town in the southwestern corner of Missouri. My mother grew up in St. Louis just a few blocks from Route 66. One of her favorite childhood memories is when, in 1955, her family of six (she was the oldest of four siblings), drove Route 66 from St. Louis to Los Angeles so that they could attend the opening of Disneyland, where she got to meet Walt Disney. This cross-country trip in an iconic station wagon on an iconic highway to meet an iconic figure is almost the height of Americana.

Federal highways helped families of modest means like my mother's to take exotic cross-country vacations, but road travel over long distances was still difficult, and the quality and capacity of highways were still inconsistent and limited in many places, especially in poorer or rural locations. General Dwight D. Eisenhower learned hard lessons about the difficulty of moving troops through the mud of the South's unpaved travel routes when he was a younger officer, and that experience stayed with him forever.[11] As president, after emerging from World War II as a war hero, he made improving the nation's ability to respond to a security threat one of his top priorities. His memories of getting bogged down on impassable roads motivated him to create a better national network of highways. His intent was primarily to make it easier for troops, tanks, and equipment to move around the nation to defend against an attack, but the side benefit was a big boost to personal automobile use and car trips.

The Federal-Aid Highway Act was signed by President Eisenhower in 1956, one year after my mother's epic trip. In the nation's largest infrastructure project up to that point, the bill used federal funds to construct a 41,000-mile interconnected system of "Interstate and Defense Highways." Its terrible context has been mostly forgotten, but one motivation for the web of highways was that cities could be evacuated quickly in the event of a nuclear attack on a major metropolitan area.[12]

The arrival of federally funded highways, along with a wide array of affordable cars and accessible gasoline, converged with the ongoing postwar recovery and American economic boom to make the allure of driving irresistible. So Americans drove. And in parallel, trains lost customers who preferred the flexibility and independence of driving their own automobiles.

Car manufacturing became one of the nation's defining industries, and oil became the world's most important form of energy. Halberstam went on to note that "the American Century was the same thing as the Oil Century—an era in which the economy was driven by oil instead of coal and in which, for the first time, the worker became a consumer as well." In the oil age, the worker could own a car and a house in the suburbs and wasn't trapped in the factories as workers in the coal age had been.[13] Auto factory workers were free to roam and could afford to do so in a way that coal miners had not been able to. As a consequence, car ownership and the number of miles driven increased steadily for decades and show no signs of slowing down. By the twenty-first century, Americans were driving more than 3 trillion miles annually in cars alone.

## JET ENGINES AND PLANE TRAVEL

The internal combustion engine operating on high-performance petroleum-based fuels opened up land-based travel at higher speeds, effectively shrinking the continent. Trips that would have taken weeks or months only required days. It changed the pattern of daily life and the layout of cities. But jet fuels and rocket fuels changed things yet again, taking humans into the air and space to cross even greater gulfs in even less time.

Air and space travel had been enticing for millennia. The story of Icarus flying toward the sun before crashing down to earth was a well-known fable about overreaching. Greek mythology presents the young god Hermes, whose winged feet implied a superhuman ability to fly. For most of history, these examples included, our fascination with flight was a purely hypothetical one. But a shift occurred when Leonardo da Vinci experimented with flying machines near the start of the sixteenth century. Subsequently, Jules Verne wrote his novel *From Earth to the Moon* in 1865, which conceived of a gun that would shoot a rocket into space.

But flight was a difficult technical puzzle to solve. For centuries, we simply tried to make vehicles that were lighter than air, such as hot air balloons, blimps, or dirigibles that could be filled with light gas such as hydrogen, helium, or heated air (using fuels like propane in the process). The first hot air balloon flight was conducted in 1783 in Paris before a series of more daring hot air balloon milestones were met: first manned flight, first crossing of the English Channel, and so forth. While these aircraft worked, they were also bulky, slow, and difficult to control.

It was not until the twentieth century that inventors realized that the key to flight wasn't to make aircraft light, but fast. By virtue of the air moving quickly over a curved wing, you can achieve the lift needed to raise the mass into the air. This is the approach used by birds, who are heavier than air but use their light, curved wings and high-speed movement to get the lift they need. The great enabler for heavier-than-air flight was the arrival of modern fossil fuels, with their high-energy density that could drive a lot of power to propellers without requiring much weight or much space. The Wright brothers conducted their first successful test flights in late December 1903 using a custom-built gasoline engine to operate their propellers. The oil age begat the age of flight.

And unlike trains, which could not go very far until rails were laid, planes did not need a sprawling infrastructure. Once you were airborne, you could go almost anywhere. Flying opened up new paths to different destinations. The idea that oil liberated humans from the ground and let us take flight is embedded in the Pegasus logo of the Mobil Oil corporation (formerly known as Standard Oil Company of New York), whose winged horse evokes the magical possibility that a flightless animal—humans come to mind—could fly.

Despite seeming revolutionary, planes were using the same technology as automobiles: a piston-driven internal combustion engine operating on gasoline. Instead of rotating wheels, the engine spun propellers. These planes gave remarkable speed compared to the relatively snail-like automobiles crawling along the ground below them. But burning the fuels in the engine requires sufficient oxygen to enable combustion. As a consequence, propeller-driven planes lost power at higher altitudes, where the air was thinner. These planes were not suitable for flying above about 10,000 feet in altitude because both the engine and pilot would struggle for oxygen.

The arrival of jet engines based on a gas turbine let planes fly faster, higher, and farther. The turbine design still relied on high-performance fuels—in this case, jet fuels that are similar to diesel—and a design that compressed air at the front end of the engine so the plane could operate at a height of 40,000 feet or more, despite the thin air. Going to higher speeds allowed flight at higher altitudes, where air resistance was lower, giving the jet planes vastly improved range. The compressor also made it easy to pressurize the cabin, allowing the pilots and passengers to breathe easily without oxygen masks at those lofty heights.

Frank Whittle, a British engineer, developed the first turbine-based jet engine. He conceived it in 1929 as a twenty-two-year-old cadet in England's Royal Air Force, applied for a patent in 1930, and spent the next decade-plus bringing it to fruition before successfully demonstrating it in a test flight on May 15,

1941.[14] The turbine engine generated a remarkable amount of power in a small footprint.

Ironically, as Whittle was still developing his jet engine, the National Academy of Sciences—a body of some of the most esteemed scientists in the United States—published a treatise claiming that the approach would never work. Their Committee on Gas Turbines published a report on June 10, 1940, that said the gas turbine engine was infeasible and "beyond the realm of possibility with existing materials."[15] Whittle proved them wrong less than a year later, and wryly noted with his own handwritten remark on a copy of the critical report, "Good thing I was too stupid to know this." Famed science writer Arthur C. Clarke once claimed, "If an elderly but distinguished scientist says that something is possible, he is almost certainly right; but if he says that it is impossible, he is very probably wrong."[16] It's as if certain members of the National Academy of Sciences, with all their prestige, were trying to illustrate the truth of Clarke's observation.

Whittle's designs—including the incorporation of an afterburner to give more thrust—are still the basis for most major civilian airplanes and fighter jets. Because the gas turbine underpins the modern air transportation system and is a major contributor to the power sector, it is one of the most important inventions of the twentieth century.[17]

As air transportation grew in importance, the desire for faster, higher planes grew alongside. There were competitions to cross various oceans or to break the sound barrier. The Concorde, a supersonic plane that traveled more than 1,300 miles per hour, served civilians from 1976 through 2003, letting them fly from New York to London or Paris in half the time of a conventional flight.

My undergraduate degree was in aerospace engineering, so planes were a part of our standard educational repertoire. As part of my studies, I held two internships in the summers of 1992 and 1993 at NASA's Ames Research Lab in the San Francisco Bay area. My student colleagues and I proudly wore T-shirts that read, "Actually, yes, I am a rocket scientist." I worked on the National Aerospace Plane (NASP), that was colloquially known as the Orient Express because that's how President Ronald Reagan referred to it when he announced the program during a State of the Union speech in 1986.[18] The goal of the plane was that it could take off horizontally from an airport on the East Coast, reach the edge of space, and then arrive in Tokyo, where it would land horizontally after a flight less than two hours in duration. The plane had a novel engine configuration with zero moving parts that used the shock wave created by the plane's high-speed travel to compress air to a high pressure and temperature before injecting fuel at high speed that would burn and heat the gases further. It was called a supersonic combustion ramjet, or "scramjet." As an aspiring engineer, I was thrilled to be part of it. The project was killed the following year, dashing my hopes. Nevertheless, the desire for faster, farther transportation lives on through space tourism programs and ongoing innovation in plane travel.

## ELECTRIFIED TRANSPORTATION

Transportation can be powered by electric motors as well as by external (steam) and internal combustion engines.* While the

---

* By engineering convention, engines are air-breathing, whereas motors are not, though both give mechanical force to move objects. Thus, the systems that burned kerosene, diesel, or gasoline are known as steam engines or internal combustion engines because they require air for the combustion of the fuel. However, electric drivetrains do not need air to operate, so they are called motors. Spaceships also do not require air for operation, as there is no air in the vacuum of space (they bring along their own oxidizer such as

idea of electric vehicles seems like a modern concept, they have been around for over a century and have intellectual roots that go back further. James Prescott Joule, whose studies in thermodynamics and fundamental physics were consequential enough to warrant a unit of energy, the joule (J), being named in his honor, saw an electric future for transportation. In 1839, he declared, "I can hardly doubt that electro-magnetism will ultimately be substituted for steam to propel machinery."[19] The transition to personal electric vehicles has been a long time coming!

Electric motors operate in a fundamentally different way than mechanical engines. They are inherently compact and quiet, and have fewer moving parts. They provide full torque even at low speeds, whereas mechanical engines give their highest power output at a few thousand revolutions per minute (rpm). That is why mechanical engines have complicated transmissions and clutches, so that the driver can get a lot of power even when the car isn't moving or is at low speeds. Unfortunately, those additional moving parts, belts, and crankshafts are all prone to failure, making the maintenance of mechanical vehicles more expensive.

Because of the high power electric motors offer, large vehicles like trains, buses, and streetcars were natural candidates for electrification. Austin, Texas, built the world's largest dam in the 1890s to create electric power for city lighting and streetcars.

---

dinitrogen tetroxide, or $N_2O_4$, to react with the rocket fuel), so their propulsive systems are called rocket motors, even though they are burning a fuel. However, in modern parlance the meaning of the words has converged such that *motor* and *engine* are often used interchangeably and with expansive contexts. That's why it is possible to have a company named General Motors that primarily sells cars powered by engines and to have motorcars and a Motor Speedway for races of those cars, even though they aren't electric. In the technological world we now have search engines and graphic engines, despite the fact that there is no literal motion.

Oskar von Miller, a famous German academic figure who lived in Munich, supported electrification of Bavarian railways, and helped create the Walchensee hydroelectric power plant south of Munich to provide the electricity. Not only do electric vehicles have fewer moving parts that can fail, but they also do not produce fumes (because they do not have tailpipes), and they are much quieter. The gentle whir of an electric motor is much softer than the thousands of explosions per minute contained inside metal combustion engine blocks, which require muffling to comply with city noise ordinances.

Transportation systems that aren't continuously connected to electric rails or overhead wires need to bring their own energy source with them. Such vehicles need an onboard storage device and a powerblock, typically a fuel tank and an internal combustion engine. Electric cars use a battery and a motor. Hybrid vehicles have a fuel tank and a battery combined with a motor and an engine, which is one reason that hybrids sometimes cost more than their conventional counterparts. One challenge of this approach is that batteries are relatively pricey, heavy, and bulky compared with gasoline. A tank of gasoline is a simple structure that holds a lot of energy. Getting 500 miles of range from a single tank of gasoline is a pretty standard achievement for modern cars, whereas it requires significant technical advance to get 200 or more miles of range from an electric car.

Modern freight trains in the United States are diesel-electric trains. They carry diesel onboard, which is much more compact than what a battery would have to be to pull the train the same distance. The engine's sole purpose is to drive a generator to power an electric motor that drives the train's wheels. That way the train combines the energy storage benefits of diesel with the high torque and ease of control of an electric motor. In Europe and Asia, the train systems are highly electrified—and also much faster.

For buses or trash trucks, which are already heavy (and there-fore would not be hobbled by the additional weight of a bat-tery) and travel a fixed route before returning home at night, electricity is a compatible source because the batteries could charge while people are sleeping. In fact, the London Electro-bus Company launched a fleet of twenty electric buses in 1907, and they worked fine for several years before the company shut down because of financial irregularities. A little over a century later, electric buses are making a comeback, especially in Chi-nese cities that are replacing entire fleets with electric buses as a way to reduce air pollution.

A similar trend is afoot for personal automobiles. Some cities (Paris, for example) are banning diesel engines because of con-cerns about the air pollution emanating from tailpipes. Norway is offering steep incentives to consumers to support electric cars. In parallel, the cars are attractive to customers because of their quick acceleration and quiet operation. As a consequence of these converging factors, electric vehicle adoption is growing exponentially. This trend has economy-wide impacts because electrified drivetrains are more efficient than combustion-based systems.

Transportation is a major energy user, responsible for about 29 percent of total US energy consumption, mostly in the form of petroleum products burned in internal combustion engines operating with about 25 percent efficiency.[20] That means that 9 of the 12 gallons in a car's fuel tank are wasted (rejected as heat into the atmosphere) and only 3 are used for propulsion. If we replaced all 3 trillion miles per year traveled by light-duty trucks and cars operating on 25 percent efficient combustion engines with 70 percent efficient electric vehicles, the economy's over-all energy efficiency would be substantially improved and emis-sions would drop dramatically. If only wind, solar, and nuclear energy were used for the electricity to charge those vehicles, then about 1.3 billion metric tons (out of nearly 6 billion tons) of annual $CO_2$ emissions would be avoided. That means electrified transportation is not only a pathway to quieter, zippier operation but also cleaner and more efficient, too. Expanding on that point, electric vehicles get cleaner with time as natural gas, wind, and solar replace coal in the power sector, whereas combustion engines get dirtier with time as their systems degrade from normal wear and tear.

Electric transportation enabled one key innovation in mass transit: an extensive subway system. While the very first stretch of the London Underground system operated on steam locomotives that produced noxious fumes from the fuels they burned, smokeless electric trains were a much better fit for the poorly ventilated tunnels. With electrification, subway systems proliferated in the late 1800s through the early 1900s. Subways transformed cities because they facilitated the mass movement of millions of riders without taking precious real estate or farmland on the surface. One observer noted that the subway essentially made New York City what it is, by bringing rich and poor people of many races and backgrounds together.[21] With densification came a need for mass transit, which enabled more densification.

Compared with other transit hubs, subways are a different kind of beast. Generally speaking, people don't want to live near airports because they are noisy, generate pollution, and attract traffic jams. But people like to live near subways because the noise and fumes are out of sight and out of mind, and they offer great convenience for moving around a city. That's the result of electrification of mass transit, which created a vast underground ballet of coordinated movements of people and machines.

Though the subway is well over a century old, it is a precursor to other concepts for underground transportation that moves people and goods at high speeds. I had the opportunity to work

at the RAND Corporation, the nation's oldest and most distinguished think tank, from 2004 to 2006. RAND was and is a
special place where many good ideas flourish. RAND employees
conceived of the communications satellite, the copay for health
insurance, and the control deck for the spaceship *Enterprise* in
*Star Trek*. I would occasionally thumb through their old reports
because I was amazed at all the gems I would find.

One breakthrough report from 1972, "The Very High Speed
Transit System" by Robert M. Salter, was particularly pre-
scient.[22] It laid out the concept of a high-speed, low-pollution
alternative to air travel that used underground tunnels with
high-powered pneumatic devices. It was inspired by the desire
to save energy but also would avoid weather-related problems
above ground. Imagine high-speed guinea pigs in their habitrails
or the tubes at drive-through banks that use air to push or pull
containers of your money from your car to the teller and back.
That's the concept presented in the RAND report. It's also an
ancestor of the core idea presented as Hyperloop by Elon Musk.
It's as if there are no new ideas under the sun.

## CHALLENGES

The good news is that transportation has been a transformative force for cities and individual lives, but it has also posed
serious downsides: concerns about energy imports, noise, lost
public space, and air pollution. For example, researchers determined that when German public transport workers go on strike,
car traffic increases, which leads to increased hospitalizations of
young children and the elderly for respiratory problems, demonstrating the comparative cleanliness of mass transit versus individual automobile ownership.[23] US roads and bridges are poorly
maintained, and their limited capacity promotes traffic congestion. In the United States, transportation is the leading source

of carbon dioxide emissions, and vehicle accidents kill more than 30,000 people every year.

The title of a prominent op-ed in 2018 in the *New York Times* declared simply, "Cars Are Ruining Our Cities," voicing exasperation that something that started out as a liberator for society has become its own form of imprisonment.[24] We have essentially given our public spaces and daily rhythms over to cars instead of people. Cars changed the layout of our urban areas, divided our cities, and promoted the rise of far-flung suburbs. They also helped foster the phenomenon of major cities whose downtowns are empty at night after everyone goes home. Cities with the most successful downtowns tend to be the ones built up before World War II—that is, before the explosion of cars. Clogged cities and highways are the most visible signs of an overused transportation system, but the invisible pollution's lingering effects are possibly even more important. That means we need to find a way to fix our transportation systems.

That op-ed went on to explain, "Both the public and a few of our bolder political leaders are waking up to the reality that we simply cannot keep jamming more cars into our cities." Unfortunately—or fortunately—just building more roads is not the way to solve our traffic problems. After many decades of building highways or expanding their capacity, only to witness more cars show up to fill the space, highway planners are starting to learn this lesson. A November 2015 article in CityLab blared "California's Department of Transportation Admits That More Roads Mean More Traffic."[25] Tongue-in-cheek commentators note that building wider roads to solve traffic congestion is a little bit like loosening our belts to solve obesity.

Thankfully, there are many other solutions. Shifting freight to rail. Mobility services. Mass transit, mass customized transit, or micro transit. Walkable cities. Bike lanes. Autonomous vehicles. Working from home. Congestion pricing.

This last option is especially interesting. When traffic volumes are high, city centers can charge a fee to drive into them. That makes it more expensive for a car to drive toward the traffic jam, giving the driver an incentive to find another route. Congestion pricing has been implemented in city centers in London and Stockholm since the first decade of the twenty-first century. Using an automated license plate scanning technology, Stockholm implemented congestion charges of up to nearly $3 per vehicle entering the city center on a trial basis starting in 2006 and permanently in 2007.[26] Nights, weekends, holidays, and July had no congestion charges. The revenues collected by the cities can fund other transportation modes, like mass transit and bike lanes, which give drivers options to get out of their cars.

By alleviating the traffic jams, congestion pricing helps address lost time commuting, fuel costs, and missed business activity. Congestion pricing in Sweden not only reduced traffic by 20–25 percent in the city center, but also reduced pollution 5–15 percent, which reduced the rate of asthma attacks among young children measurably.[27] (Children are more affected by air pollution because their lungs are still forming, they are outside more, and they engage in more physical activity than adults.) These examples demonstrate the power of markets to address social problems.

## A NEW AGE FOR RAIL

One of the simplest and most powerful ways to address transportation's energy challenges is to reinvest in freight railroads.[28] Freight transport is a much bigger part of the nation's infrastructure than most people realize. In 2013, about $50 billion worth of goods was transported as freight cargo every single day in the United States alone.[29] Trucks move 29 percent of the freight ton-miles, but are responsible for 77 percent of the sector's emissions.

Astonishingly, empty trucks account for about one-fifth of the truck miles traveled. Between the rise of Walmart with its truck-based logistical system and the spread of internet-based retailers such as Amazon, highway freight tonnage grew by 45 percent between 2000 and 2014. According to the US Department of Transportation, the existing population of trucks on congested highways already substantially impedes interstate commerce, and projections suggest highway congestion will get much worse in the coming decades.

Mass transit for commuters and high-speed passenger rail get a lot of attention and capture the imagination, but it is the more mundane movement of goods that presents a worthwhile opportunity for system-wide improvement.

Trucks are convenient because they enable flexible point-to-point operation, but they are relatively inefficient, dirty, dangerous, and destructive to our roads. Rather than wait for some still-unrealized technological breakthrough, we could instead expand our national freight rail system. At first blush, it would be easy to believe that freight rail's day has passed. From its peak a century ago at more than 250,000 miles, today there are fewer than 95,000 miles of track for Class I railroads, as rail lost market share for the movement of people and goods to air travel and the interstate highway system.[30] From 1990 to 2013 alone, the US population increased 28.2 percent while track miles decreased 28.6 percent, despite increases in shipping and freight movement.[31]

That decline didn't just happen. It was a policy choice carried out over decades. The decision to invest trillions of dollars in the interstate highway system was a vote for trucks over rail. Today, the national freight transportation infrastructure has about $6 trillion in assets, with more than half that total locked up in highways. In contrast with private-sector trucks that operate over public highways (making the interstate highway system essentially a subsidy by taxpayers for trucking companies), the freight railroads are almost entirely private.[32]

Since revenues bottomed out in the 1970s, financial restructuring enabled the railroads to invest in better efficiency. Consequently, revenues and profit have risen for decades despite the decreased trackage available.[33] American railroads have become extremely efficient and productive, moving increasing volumes of freight over a shrinking infrastructure.[34]

Rail moves 40 percent of freight as measured in ton-miles but is responsible for only 8 percent of freight transportation carbon emissions.[35] Even though both trucks and locomotives use the same fuel—diesel—railways emit less $CO_2$ per ton-mile of freight movement because rail is much more energy-efficient than trucking. By one estimate, moving freight by rail instead of trucks could save up to 1,000 gallons of fuel per carload.[36]

And freight rail has the potential to get cleaner more quickly than trucks, ships, or planes. According to the Bureau of Transportation Statistics, the median age of the 25,000 locomotives in the United States is less than thirteen years, so natural fleet turnover patterns offer a chance to bring in newer, cleaner versions.[37] That means the rail fleet can be cleaned through investments in just tens of thousands of locomotives, compared with more than 10 million heavy-duty trucks on the roads today. For instance, switching locomotives to compressed natural gas could reduce emissions while reducing imports of crude oil.

To be sure, trucking is great for the last few miles when delivering goods. That capability is useful, as most of us do not live next to train tracks. But more than two-thirds of freight travels more than 500 miles. Rail transportation is so efficient that even if cargo must travel a longer distance by rail than it would by point-to-point trucking, shipping by rail still uses far less energy.[38]

Transferring freight cargo by rail instead of by truck is also safer. Freight transportation is responsible for approximately 100,000 injuries and 4,500 fatalities each year, and trucks are responsible for 95 percent of those injuries and 88 percent of those deaths.[39] Most of the people killed by trucks are in passenger vehicles sharing the road with 18-wheelers. A study in 2013 concluded that the additional risk of fatalities from heavy trucks is equivalent to a gas tax of $0.97 per gallon.[40]

By contrast, rail transportation is responsible for about 4,000 injuries and 500 fatalities, the vast preponderance of which were from trespassers on the railroad right-of-way.[41]

Not only will roads be safer, but they will be in better condition, too. The Highway Trust Fund that provides money for maintenance and repairs goes broke every year because revenues (collected via a tax on gasoline and diesel fuel) haven't kept up with expenses. One way to provide the trust fund with enough money is to raise the gasoline tax, set at 18.4 cents per gallon in 1993. Another way is to reduce the wear and tear by removing the heaviest vehicles. Damage to roads scales with axle weight to the third power. So one 40-ton truck causes more than a thousand times the damage of a typical 4,000-pound car. As one seminal study noted, "For all practical purposes, structural damage to roads is caused by trucks and buses, not by cars."[42]

Until recently, the freight railway network was being stressed on many routes by hauling coal from fields in the western United States to power plants around the country. Coal trains are huge: 100 hopper cars that can each carry 100 tons, all pulled by six locomotives, each with 3,000 horsepower. In terms of ton-miles, coal still comprises the single largest commodity moved by the freight rail system. Coal's decline in the power sector as it is displaced by cheaper, cleaner natural gas, wind, and solar power opens up spare capacity in the rail system that could be used for moving other goods.[43]

Those railroad right-of-ways from the western coal fields to the East also have potential as routes for alternative energy. Following up on the old idea of lining train tracks with telephone poles, we could couple rail lines with a national high-voltage direct current transmission network, spanning the heart of the windy Great Plains and sunny Southwest, thereby enabling better integration of renewables, cleaning up the power sector further.[44] We could even put those power lines underground to reduce their vulnerability to windstorms. Since the best solar and wind resources are often far from major demand centers like large cities, developing a national grid that can easily move power around would be advantageous.

Not only would a nationally coupled rail and power-line network reduce emissions and spawn more rural economic development, but it would also improve grid reliability. Laying electricity along the tracks also opens up the door for electrified freight trains. Such trains are common in Europe, and it may be simpler to electrify freight rail transportation than to build out the charging infrastructure for electric road vehicles.

The rail system sprawls across the continent, so many of the economic benefits from a return to freight rail would accrue to rural areas. In addition, locomotive engineers make about 30 percent more per hour than truck drivers.[45] Even the other jobs—rail yard engineers, signal and switch operators, conductors, and so forth—pay more than driving trucks. Those higher wages would have economic benefits that rippled out. Seeking job growth for these roles is an enticing idea.

One simple way to encourage the switch from road to rail is to put a price on carbon. A carbon tax would harness the efficiency of markets while sending a price signal that rewards the more energy-efficient and cleaner option of rail transportation. Another approach—one that wouldn't also put motorists in the crosshairs—would be to raise money for road maintenance via a fee based on miles driven and vehicle weight.[46] That would target the vehicles that do the most damage and stop the subsidy of heavy trucks by the drivers of small personal cars. By more closely aligning the costs with the damage, trucking would lose some of its competitive advantage compared with rail.

A carbon price and update to our gas tax model would likely encourage a lot of switching to rail for freight, but increasing throughput (ton-miles) on rail without other improvements could degrade other key performance metrics such as delivery time and reliability. Since many freight customers are very sensitive to those factors, commensurate investments have to be made in optimizing performance, double-tracking where possible, adding new tracks, and alleviating bottlenecks.

Expanding track miles is an obvious step forward, though not the only one. Adding more sidings or double-tracking at congested zones can facilitate the operation of more trains in different directions and allow trains operating at different speeds to more easily share the same track.

But just laying a lot more track isn't enough. As a pair of major studies by the RAND Corporation in 2008 and 2009 noted, increasing the national freight rail capacity will require a variety of strategies beyond direct infrastructure investments. Such measures include revising regulations, flexible pricing, deploying new technology, and implementing improved operating practices. For instance, operational enhancements to more efficiently use existing tracks might be just as important as building more miles of track, but those changes need to be informed by more detailed and extensive modeling to identify locations of bottlenecks and to develop schemes that avoid them.

Another way to increase capacity while cleaning up the transportation sector is to enlarge and improve the fleet of locomotives. Incentives for rail companies to buy newer, cleaner, more efficient locomotives would simultaneously clean up and expand capacity. Partly because of this potential, the International Energy Agency noted that the rail sector can play a key role in reducing global $CO_2$ emissions.[47] More routine and detailed inspections of rail systems can also improve safety and throughput by allowing heavier freight loads and faster train movement.

In the end an old idea—moving goods by rail—might be the modern innovation we need to reduce energy consumption and avoid $CO_2$ emissions while making roads less congested, safer, and more enjoyable for motorists.

## SMART MOBILITY

In the upcoming discussion on cities in Chapter 5, it will be noted that reducing waste is a key pathway to making a city smart, and making a city smart is a key enabler for reducing waste. This same idea applies to transportation: reducing wasted time and space in transportation will also improve cities.

In fact, better transportation may give urbanites their first glimpse of a smart city's benefits, by reducing physical waste and turning wasted time into a resource.[48] Reducing the footprint of transportation means cleaning up the fuels, making the vehicles more efficient, reducing trip distances and duration, increasing vehicle occupancy, and reducing the number of trips.[49] If people live close to work, they can walk or bike, or use mass transit if population density is high enough to support it. Scientists have concluded that building protected bicycle lanes leads to dramatically increased ridership.[50] Because bicycles require so little space compared with cars, they can reduce congestion on the roads.

Turning to electric vehicles and switching from automobile ownership to ride-sharing in self-driving vehicles could reduce congestion by smartly controlling traffic flows using knowledge of where other travelers are going. Also, cars could be smaller. Most of us drive a single-occupancy car that has room to seat four or five even though we rarely need it. With point-to-point mobility services for our commuting, vehicles could be tailored for the purposes of moving just a few bodies, which means they could be smaller. Our recent research at the University of Texas at Austin has concluded that when the full life-cycle costs of owning your own vehicle are considered (the cost of insurance and taxes on your garage at home, paying to park at work, maintenance, fuel, and the lost productivity of time spent driving), using a mobility service is the best economic option for over a quarter of the population using standard conditions from 2017.[51] As the prices for mobility as a service (MAAS) drop, then that will be the economic option for a much larger fraction of society. Professionals who live in suburbs would benefit from using mobility services: instead of wasting their time driving, commuters can rest, read emails, place phone calls, or conduct other business. That work can create economic value—and reduce workers' office hours so they can get home earlier for dinner.

By some accounts, approximately 30 percent of the cars in congested urban traffic at peak times are looking for a place to park.[52] British drivers on average spend four days per year looking for open parking spots, which raises their stress levels.[53] If drivers can get access to real-time pricing for parking spaces and maps that show where spots are available, they can spend less time hunting and go straight to the spot they want. If cities promote shared or autonomous cars in constant motion instead of point-to-point private cars that are parked at both ends, they can reduce the number of parking spaces dramatically, opening up space that can be dedicated to buses or bikes, reducing congestion further. Plus, we may no longer need so many parking garages, freeing more space for office buildings or residences, amplifying density further. Researchers at the Center for Transportation Research at the University of Texas at Austin used sophisticated agent-based models to determine that shared autonomous vehicles (SAVs) would reduce the number of cars and parking spaces needed by an order of magnitude, meaning each new SAV put into use could displace about ten other cars because it could serve so many riders.[54] Doing so would save a lot of urban space. It would also increase total miles traveled by about 11 percent because the SAVs would stay in constant motion as they move from the location where one rider is dropped off to the location where the next rider is waiting to be picked up. But despite the additional mileage, by avoiding cold starts and running more efficiently, SAVs would reduce energy usage by 12 percent, greenhouse gas emissions by more than 5 percent, and emissions of other pollutants like sulfur oxides, nitrogen oxides, carbon monoxide, and volatile organic compounds, by 18–49 percent.

Even little advances in sensing and transaction technology can improve a city's transportation functionality. After London implemented a new system in 2014 to let riders pay bus and Underground fares more quickly and safely with the touch or tap of their mobile device rather than the traditional issuance of fare cards, the administrative cost of running the payment system dropped and commuters saved time.[55] A smart signaling system in Denver, Colorado, helped synchronize trains and traffic signals to avoid the awkward situation in which a long train accidentally blocks intersections unnecessarily and to keep the train running continuously near its intended speed without long delays to wait for the traffic to clear.[56]

From my perspective, the future of mobility can't get here soon enough. Our current model of owning cars is insane. It wastes time, space, and money. On average, an American buys a $30,000 car and then uses it 4 percent of the time.[57] Then we pay to park it at home—with a garage that takes up valuable

space. At the same time, more people are using mobility services, and the percentage of teens getting their driver's license
continues to decline. Smart students and post-docs who work
with me have wondered out loud whether most Americans
should stop owning cars.[58]

On its face, spending so much money for an appliance that
starts losing value immediately, takes up vast amount of our free
time, and is rarely used seems ridiculous. Working with a really
smart undergraduate student and senior researcher (Gordon Tsai
and Dr. Todd Davidson), we conducted an analysis of the all-in
cost of car ownership. The costs of traditional car ownership go
far beyond the price tag, including interest paid on car loans,
insurance, taxes, fuel, and maintenance. Some expenses are
not obvious, such as parking, property taxes, and construction
costs for home garages, and the value of our time. The average
American spends 335 hours annually behind the wheel driving
over 13,000 miles.[59] The full costs of car ownership—including
car payments (>$5,700/year), maintenance (>$1,100/year), fuel
(>$1,500/year), parking (>$1,400/year), and lost productivity (>$6,700/year)—are nearly $17,000 per year for an average
American. We found that mobility services such as ride-hailing
and ride-sharing apps—which few people today would consider
their main mode of transportation—will likely provide a compelling economic option for a significant portion of Americans.
In fact, if the full cost of ownership is accounted for, we found
that potentially one-quarter of the entire US driving population
might be better off using ride services than owning a car.

Add in the hours we spend maintaining, cleaning, and managing our cars, and it becomes clear that America's focus on personal car ownership is costing us a significant amount of time.

How much is it worth to us to avoid the stress of driving
around town and to use that time more productively or enjoy-
ably by catching up on email, reading a book, or taking a nap?

Some professions are more suited to using time riding in a productive way: It's probably easier for a lawyer to clock billable hours while riding to work than a plumber, for example. When these costs are included, mobility services are an economically preferable option over traditional car ownership for about a quarter of typical American drivers.

The rise of autonomous vehicles used for carpooling and ride-sharing services could make mobility services even more compelling, particularly when you consider the economics from the service provider's perspective. As the price for autonomous vehicles goes down, the cost of delivering ride services drops. That means more consumers are more likely to use them, expanding the overall market. Uber, Lyft, Alphabet, and many of the automotive companies understand this. It's one of many reasons why there is an epic race to capture market share and eventually be the first to deliver fully autonomous vehicles.

Some trends do appear to be working in favor of increased use of mobility services, despite the cultural importance of car ownership. As the United States, and the world more broadly, continues to adopt ever-greater levels of digital communication, more people will be able to complete work while on the go. And the increase in people moving to cities has resulted in denser urban centers, increasing traffic congestion and making the case for alternatives to traditional car ownership.

Even changes in how different generations consume goods and services might be playing a role. Millennials have shown tepid interest in following in the footsteps of prior generations when it comes to car ownership. It will be interesting to observe whether Generation Z shows more desire for cars and the suburbs as they begin to enter parenthood and pursue affordable housing.

In addition to common mobility services today, Uber and Lyft might soon be joined in force by micro-transit operators like Ford's Chariot shuttle service. As mobility services become more mature, we will likely see those services become even more convenient by meeting specific needs such as transporting youth, the elderly, or disabled people and even assisting in disaster recovery efforts. The increased level of service could create a virtuous cycle that reinforces the value of mobility services, producing greater adoption, which further lowers costs and leads to even greater adoption. When all's said and done, the ease and economic benefits mean that the transition to mobility services might take place faster than we think.

The rise of autonomous vehicles could accelerate this transition. Because autonomous vehicles eliminate the drivers, who are expensive, they will lower the cost of mobility services further. It has been estimated that autonomous taxis will cost $0.35 per mile, much cheaper than a taxi ride today, which is usually a few dollars for each mile.[60]

In addition to taxis, trucking might be another sector where self-driving vehicles take off. There are a variety of reasons autonomous operation is appealing for long-haul freight, namely that stops for the drivers can be avoided, which improves delivery times. And routes can easily be prescribed ahead of time on major highways. A *Financial Times* article wryly noted that while there is a lot of political hand-wringing about the job losses truckers face from autonomous operation, it is actually the other way around: the lack of truckers might accelerate the adoption of autonomous trucking.[61] Truckers must pay $5,000 to $10,000 out of their own pockets for training and certification, which is a major obstacle for prospective employees. At the same time, those same workers can find more attractive jobs in construction or other industries with similar pay and benefits but without the upfront costs. Consequently there is a shortage of truckers, which could accelerate the use of advanced technologies to fill that need.

The rise of mobility services and self-driving cars might be as transformative as cars were in the first place. Imagine this scenario: We all have our own chauffeur who picks us up at our door the minute we're ready and drops us off at work or the grocery store, driving along roads with smooth traffic and sparing us the hassle and time of finding a parking spot. En route, we can read, text, think, sleep, or talk on the phone without fear of causing an accident.

While this vision seems incompatible with "American car culture," from what I can see, the culture is ready for the change. While the driver's license was a threshold of adulthood for me and my peers when we came of age, many teens today think differently. Many people cluck that they are worried texting will interfere with teens' driving, but from teens' perspectives it is the other way around: the driving is interfering with their texting. And it isn't just the kids: Look around the next time you're stopped at an intersection. Most of the adult drivers in the cars next to you will be looking at their phones. Most of them look as if they'd be happier with their eyes off the road.

It might be a private car and private chauffeur if we could afford it, or there could be a few others in the same vehicle if we wanted to save costs by carpooling. Regardless, the service is still door-to-door. Now imagine the chauffeur is a robot or hidden microprocessor.

These mobility services with autonomous vehicles will save energy in a number of ways. Robotic drivers will be programmed to follow the best practices of driving—they won't have lead feet and bad habits. Embed more information into the cars and the surrounding infrastructure, and traffic will move more smoothly, reducing congestion, smog, and energy consumption. A suite of connected cars that know what the other cars plan to do will make traffic lights obsolete; instead, the cars will continuously weave around each other at crossings. Safety will improve because each car will automatically know where the others are headed, reducing the risk of collision, just as planes do in the sky.

And since the cars will be better matched to the needs of the riders, there won't be lone commuters in gas-guzzling SUVs that only make sense on the occasional weekend. When you need to tow a boat, you could arrange for a robot-driven truck. Otherwise a smaller commuter car will be the primary vehicle of choice.

The cost of the vehicles will be shared through the mobility service company, keeping ownership costs down per mile traveled. Rather than each of us paying 100 percent for a $30,000 car we use 4 percent of the time, we will all pay for a more expensive car, but only when we need it, with one car meeting the needs of many. Auto insurance companies will likely have lower rates to reward the improved safety of chauffeured cars compared to cars we drive ourselves, creating a nice market incentive to get a ride. People who really want to drive will pay extra insurance to reflect the additional risk they are introducing on the roads. Parents who really want their teenagers to drive can pay a premium yet higher. Once these technologies and market signals align to point the same direction, the trends will be irreversible.

If you love to drive and worry that society will lose this critical skill set as we hand our transportation needs over to machines, then consider this: some clever entrepreneur will sell you the opportunity to drive old beat-up cars in a circle on a dusty ranch while reliving the past. After all, that's what we do when we want to teach our kids how to ride horses.

## Chapter 4

# WEALTH

Once energy investments into agriculture increased food production, society could accommodate specialized labor. Instead of people foraging, hunting, or gathering food to feed themselves, a few percent of the population would work as farmers and could feed the rest of the population. By developing agriculture, which allowed groups to stay put instead of wandering for food, and by specializing in different activities, it was possible to accumulate wealth. Then, millennia later, when additional energy inputs allowed for significant productivity gains to agriculture, each farmer and each acre could feed more people. That allowed for even greater specialization of labor and accumulation of wealth. Once people had wealth, they migrated to cities, which were fed by a transportation system that brought agricultural products from farm to table. Using energy to increase food productivity is a key enabling step of that whole process.

Today there are over 500 billionaires and thousands of millionaires in the United States. This trend of spreading wealth, in contrast with aristocracies or monarchies, where wealth was concentrated into the hands of the very few, started in the late

1800s and early 1900s with some very recognizable names: Rockefeller, Carnegie, Vanderbilt, Stanford, and Ford. They built fortunes on energy resources and industries dependent on energy,
such as steel, railroads, automobiles, and shipping. Widespread
access to energy after World War II in industrialized countries
resulted in a broader democratization of wealth that lifted hundreds of millions out of poverty and vastly expanded the global
middle class. Modern wealth is enabled through access to some
form of energy, and those without energy remain poor.

## ENERGY, LIGHTING, EDUCATION, AND WEALTH

If water is life, energy is quality of life. Modern energy—
electricity, natural gas, or propane instead of cow dung, straw,
peat, or firewood—lets households get out of the cold and cook
their food in a way that produces less smoke, soot, and ash. Importantly, access to energy enables access to education. And
education is one of the most important pathways to affluence.

In urban areas connected to a natural gas distribution system in the 1800s, such as London and Paris, residents would
have natural gas lighting indoors that they could operate with
a turn of a key at the wall that would open or close the flow of
gas. While the availability of gas lighting was transformative, it
still produced fumes, heated the rooms, posed a risk of carbon
monoxide poisoning, and occasionally started fires. The arrival
of electric lighting was a substantial improvement because it
addressed those issues.

Modern energy systems also enable modern sanitation systems, opening up the possibility of indoor plumbing. In many
countries where indoor plumbing is not ubiquitous, taking care
of normal body functions can be an embarrassing or tedious process. In countries like India, open defecation is very common.
As a consequence of the lack of modern sanitation and toilets,

many female students in India stop going to school when they hit puberty. Because they don't have private facilities to take care of their personal hygiene needs, they simply stay home.[1] With modern energy systems that enable schools to install modern plumbing and sanitation systems, the educational prospects for students—especially females—are improved. As a result, their economic opportunities also improve.

One notable ad titled "Keep the Boy in School" from Case Tractors in 1921 extolled the importance of energy as a pathway for education.[2] The ad pleads with farming parents to buy a tractor powered by kerosene as a way to free up their son for his education: "With the help of a Case Kerosene Tractor it is possible for one man to do more work, in a given time, than a good man and an industrious boy, together, working with horses." It closes with "Keep the boy in school—and let a Case Kerosene Tractor take his place in the field."

Perhaps it isn't a surprise that leading business figures who got rich from energy invested in education. Most notably, John D. Rockefeller, father of Standard Oil and for a time the world's richest man, sponsored the creation of universities, including Spelman College (his wife's maiden name was Spelman) and the University of Chicago. Other universities—Stanford and Carnegie Mellon come to mind—bear the names of people whose fortunes were tied either directly or indirectly to the energy industry.

The connections between energy and wealth are more nuanced than people expect. For people in poverty, access to energy improves their economic fortunes. Not only can that energy be used to improve their educational opportunities, but it can also be used to operate a business, build a factory, and make goods. More energy means more economic opportunity. And in the United States, people can make money from the raw resources on their land, because they are paid a royalty for energy production that takes place on their acreage.

But it is also true that rich people consume more energy. As people get richer, they tend to consume more meat (which is energy-intensive), more electricity for climate control and appliances at home, and more gasoline for transportation. Consuming energy makes us rich, and getting rich makes us consume energy. Generally speaking, there is a linear relationship between wealth and energy consumption: the richer a country is, the more energy it consumes per capita. And the more energy it consumes per capita, the richer it gets. Wealth tends to put people and countries on an energy treadmill.

Of course, there is always more to the story. Cultural choices, policies, and other factors affect energy consumption. European countries are about as rich per person as the United States or Canada, but their residents consume about half as much energy per person. Part of that is because the United States and Canada are large countries with low population densities, both are major energy producers, and both are relatively young. Smaller, older countries in Europe that do not produce much energy will be different in multiple ways. Because they are smaller, they spend less energy on transportation. And because they have higher population densities, they tend to live in compact urban areas with smaller homes that are more efficient to operate. Also, because they are less self-reliant on energy, saving energy is an important cultural and policy priority to reduce the security implications of energy imports. In addition, the countries can often trace their roots back centuries or millennia. As a result of this longevity, they tend to take a longer-term view of the future and worry more than the United States about the impacts from climate change, and they have stricter policies related to energy efficiency that mandate efficient automobiles and appliances, and taxes that drive prices higher to use market signals to reduce consumption. In the end, those combinations of factors achieve lower energy consumption per person without significantly hampering economic growth.

Other countries, such as China and India, have lower energy consumption overall, not because of policies but because of poverty. Energy access is low and many people do not necessarily have the money to afford it when it is available. Ultimately, a typical global citizen consumes about 75 million BTUs of energy each year. A typical resident of the United Kingdom consumes twice as much, and an American consumes nearly twice as much again. Some countries' residents consume even more energy. Icelandic residents consume a lot of energy because it's cold and a lot of heating is needed and because geothermal energy—that is, energy from the heat of the Earth—is cheap and abundant in Iceland. In fact, it's one of the country's defining features. Iceland uses that energy to make electricity and to manufacture aluminum and high-value products, which means they are able to convert their abundant energy resources into financial gain. Residents of Bahrain also consume a lot of energy because they have a lot of energy, which makes it cheap locally, and because they need a lot of air conditioning in their hot climate. Russia and Saudi Arabia consume a lot of energy per capita but are not very wealthy. That is because they use the energy mostly to make heat for heavy industries such as metalworking or refining rather than for higher-value products such as chemicals.

That's another piece of the energy and wealth story: the form of energy matters. There is a good—but not great—correlation between energy consumption and wealth, but there is an even better correlation between *electricity* consumption and wealth. Countries that use their energy for cruder applications, such as heating and cooking, are not as wealthy as those that use their energy for sophisticated purposes, such as generating electricity.

Electricity enables high-tech industries such as software, 3D printing, and robotics, all of which are more lucrative than some of the conventional heat-based manufacturing. Richer societies prefer electricity because it's easier to control, quieter, and cleaner at the point of use and because certain appliances such as air conditioners and computers require it. And electricity is important to fostering economic growth because of its connections to electric lighting and information technologies, both of which promote upward mobility compared with traditional, manual career pathways. That means electricity consumption enables affluence, and as people become richer, they consume more electricity.

## ENERGY, FACTORIES, AND THE INDUSTRIAL REVOLUTION

Wealth has always been tied to energy, so as the forms and availability of energy changed, the nature of wealth changed as well. Until the mid-1800s, wealth was primarily tied to land, and that land was used for producing goods that were usually agricultural in nature, such as timber, wool, leather, or food. Vast kingdoms were built on territories owned by a handful of families. But the industrial revolution allowed people to accrue wealth by industry and other activities, kicking off an ongoing era of innovation and entrepreneurship alongside economic growth. Transitioning from an agrarian economy to a modern industrialized economy occurred hand in hand with transitioning from wood, water, and wind energy to fossil fuels and electricity. As energy forms continued to advance and people gained broader access to petroleum or the products from coal, it helped to lift an entire swath of the population out of poverty, reducing inequality, increasing economic opportunity, and expanding the middle class.

By many estimates, George Washington was the richest president in US history, not because he had a lot of money, but because he had a lot of land. Land was a measure of wealth for millennia. It also takes up a lot of space, cannot be moved, and is hard to use for spending or making investments. The wealth derived from land took a lot of space to grow crops, harvest timber, or graze cattle.

By contrast, the wealth enabled by fossil fuels was different. The energy density of fossil fuels was a proxy for the ability to concentrate wealth. Once coal was available, factories could generate enormous wealth without requiring as much land as agriculture. Suddenly people could get rich without owning land and they could take their wealth with them, freeing up the opportunity for capital to flow from place to place. Those factories would not have been possible without the invention of the steam engine and the availability of coal that could be cheaply burned to make steam to operate those engines. As those engines replaced waterwheels, windmills, and muscle power, wealth grew and productivity increased.

Along a footpath above the riverbank in White River State Park in Indianapolis, Indiana, a large piece of limestone is engraved with a historical note that celebrates the revolutionary impact of steam-powered machinery. It arrived in 1864 and tripled production of the region's famous Indiana limestone. Before that, the heavy limestone was cut, lifted, and moved by hand. The new capabilities of machines running on fossil fuels to quarry and cut limestone in greater volumes and with less manual effort, combined with the rise of transportation networks such as coal-fired trains, meant that heavy limestone could be shipped hundreds of miles to customers far away. It is the same gray limestone that was used to build the Tribune Building in Chicago and the National Cathedral in Washington, D.C. New

York City was a popular destination for the beautiful rocks: Indiana limestone was used to build the Empire State Building, Grand Central Station, and the Flatiron Building. The cost to move so much heavy material from Indiana to the Eastern seaboard by horse-drawn wagon across Appalachia or by boat down the Mississippi River and around Florida is hard to fathom. This example of fossil fuel–powered steam machinery enabling conversion of local resources into higher-value products that could be shipped far away was being repeated in other locations. Coal gave rise to global centers of industry that made products for consumption far away. This kind of growth happened in Manchester, England, in the 1800s for textiles; Pittsburgh, Pennsylvania, in the 1900s for steel; and Chinese manufacturing towns today for a whole host of manufactured goods.

A few technologies powered by modern forms of energy were critical enablers for the rise of industrialism and wealth generation. The steam engine replaced water and wind power, but this movement of wealth generation from out in the fields to the interiors of factories benefited from two more inventions in the late nineteenth and early twentieth centuries: air conditioning and indoor lighting. Electric lighting that illuminated larger spaces for longer durations enabled workers' shifts to extend into the nighttime hours.[3] Today, it is not unusual for factories to operate around the clock. And air conditioning made the work more bearable for workers in hot weather.

Electricity led to more productivity in factories and the economy as a whole. It also made them safer and easier to operate. So electricity caught on quickly. In the span of thirty years at the start of the twentieth century, the electrification of Chicago's factories grew from 4 to 78 percent.[4] Although electrification displaced many workers because electric motors could do the work of many men, it did not reduce employment. Instead it fostered new enterprises and created demand for different kinds of labor.

Energy doesn't just light or cool our buildings: it also transforms raw materials into valuable products. In medieval tales, including the popular Rumpelstiltskin, magical alchemists are called upon to transform common materials such as lead or hair into gold. That process is a daily activity by the world's petrochemical industry. Common crude oil typically sells for $35 to $85 a barrel, but is transformed into paints, chemicals, plastics, lifesaving pharmaceuticals, and other commodities worth much more. The legendary perfume, Chanel No. 5, worth more than tens of thousands of dollars per gallon, is made from oil worth a few dollars per gallon. By one count, more than 95 percent of the ingredients in fragrances come from petroleum. Energy can convert a common substance (oil) that is worth about as much per pound as lead into something worth more than gold. Maybe Rumpelstiltskin was not as fantastical as we once thought.

## MINERAL RIGHTS AND WEALTH

Although land ownership is substantially less correlated with wealth than it used to be, it nevertheless has its advantages. In most parts of the world, the government, not the landowner, owns the minerals below ground. These minerals include ores, precious metals, and fossil fuels. But in eastern Canada and in the United States, the mineral rights (that is, ownership of the minerals) are also a private property that can be owned. That means when minerals, oil, or gas are extracted from the ground, a royalty is paid to the owner. When oil booms happen in the United States, not only do oil and gas companies get rich, but landowners who happened to have property in areas with accessible oil reserves could make significant sums from royalties.

That alignment between the financial interests of the producers and the landowners helped encourage exploration because wildcatters could take private risk to achieve private gain. This approach conflicts with practices in other locations around the world where the government owns the minerals but the citizens own the land. In those cases, it is private risk and public gain. That is, those whose land is at risk do not receive direct financial benefits. The landowners endure the disturbance of noise, trucks, dust, and lights, but the government gets the money. This misalignment of interests inhibits energy production from private land in most countries.

Texas is a living embodiment of this phenomenon of oil wealth. Texas, the second-largest state in the United States (second in size behind Alaska and second in population behind California), had an agrarian economy for centuries. Cotton and cattle drove the economic fortunes of the state. But in 1901, the legendary well known as Spindletop completely changed the balance of power among oil-producing nations when it began to gush hundreds of thousands of barrels of oil into the air. Texas became an oil state. This oil production made many millionaires, with the gushers a symbol for sudden wealth. Words like "strike it rich," "black gold," and "Texas tea" are references to drilling for oil and striking a reserve of oil, helping people to become rich overnight. Those with a lot of land with oil would become "big rich."[5]

This idea of the rich Texan is captured in a whole host of popular cultural depictions, even including cartoons such as Bugs Bunny, which had an episode in which Yosemite Sam plays a rich Texas oilman who is confounded by rabbit holes, among other hijinks. The 1955 movie *Giant*, starring Elizabeth Taylor,

Rock Hudson, and James Dean, also shows this. James Dean plays Jett Rink, a ne'er-do-well, hard-drinking ranch hand who strikes oil. The scene of him dancing underneath the gusher, smile splattered with oil, is one of cinematic history's most famous moments. The opening credits of the famous television series *Dallas* in the 1980s also capture this transition from an agricultural to modern energy-based economy by showing the movement from ranches to urban areas.

Texans use the idea of the "rule of capture" to govern oil and gas policy, a tradition that originated in English common law for hunters. Because the king owned the country's wildlife, it was illegal to hunt his animals. However, if you captured a king's stag on your land, you could keep it. Under the rule of capture, if you can produce oil and gas from your land, you own it. But with a big enough pump, it is possible to deplete the oil and gas underneath your land and to start draining the oil and gas from your neighbors' lands, too, as the fuel can flow underground from one spot to another. The landowner with the biggest pump wins. The 2007 movie *There Will Be Blood* grippingly captures this phenomenon. In one scene, the main character Daniel Plainview (portrayed by legendary actor Daniel Day-Lewis), gloats about drinking a neighbor's milkshake; that is, he drained his neighbor's land of oil and gas before the neighbor knew any better.

## ENERGY, TOOLS, AND INNOVATION

The history of innovation—of making more capable materials, machines, and tools—goes hand in hand with the development of more advanced forms of energy. Coal has been particularly critical over the last century and a half. In addition to providing heat to power the steam engines used in factories and trains, it is a key ingredient in the process of making cement and steel. In kilns, coal provides the heat that turns inputs like limestone, clay, and other materials into cement mix. Though concrete seems simple and old-fashioned, it is actually a symbol of modernity because of the roads, airports, power plants, and water treatment systems built from it. Large-scale production and use of concrete rose alongside the boom in coal usage.

Because steel and coal are so tightly linked, they also rose in lockstep with each other. Coal is used for steelmaking in two ways. First, it's used as a source of heat to melt iron ore so that it can be forged, cast, or otherwise shaped. Second, coal is a source of carbon for hardening steel. For older metalworking, wood or charcoal was burned to melt or soften the metal. Because there wasn't enough heat from the wood, the blacksmith would add energy by hammering to soften and shape the metal. The image of the muscular, sweating blacksmith symbolizes the pre–fossil fuel era. In some locations, waterwheels would operate bellows to improve the flow of air into furnaces to increase burn temperatures, but they were not always available. Water power and muscle power ultimately were replaced by the convenience of burning coal.

Steel is remarkable: strong yet malleable, it is a critical material for building bridges, machines, skyscrapers, and other manifestations of an advanced society. But initially steel was used for making stronger tools, such as hammers, or sharper axes. With better tools, more energy could be accessed. Better axes that were larger, sharper, and stronger raised the productivity of the timber industry. In the United States, which spread across the continent quickly from the late 1700s through the 1800s, the demand for lumber was increasing. Wood was used as a construction material and for fence posts during westward expansion. It was also the primary fuel in the United States for cooking, home heating, and industrial manufacturing.

As a result, the United States, which had 800 million acres of virgin forest in the 1600s, was rapidly deforested. As trees were cut down faster than they grew back, harvesters had to move farther west and north to get access to trees. The use of coal to make steel to make rails and locomotives powered by coal enabled timber to be harvested from more remote locations. Forests in upper midwestern states, including Minnesota and Wisconsin, supplied wood to many markets by floating or shipping felled trees on lakes and rivers to Chicago for conversion into lumber that would go back out by train for customers in the Great Plains or on the East Coast.

Timber became increasingly expensive as harvesters had to travel farther to get it and then had to transport the product back to market. But coal, which facilitated this deforestation, also solved it. Coal enabled the production of the steel that created stronger tools that could be used for tunneling and mining. Pennsylvania had abundant coal seams belowground and was close to major seams of iron ore. By using new tools, the coal could be extracted, which in turn yielded better steel tools for cheaper coal mining. Because coal had twice the energy density of wood and produced less smoke, ash, and odors, and was closer to market (in Appalachia, not the upper Midwest), it quickly became a preferred source of fuel.

Coal also made steel drill bits that enabled oil drilling, which started in earnest in 1859, also in Pennsylvania. Oil had been used for millennia for oil lamps, for tar, and as an ointment; it had typically been collected from surface seeps like those in the Middle East or at the La Brea Tar Pits in Los Angeles. But with new tools and a hunch, Colonel Edwin Drake was the first to intentionally drill to extract oil from below an oil spring outside Titusville, Pennsylvania. His partner was a salt well driller. Initially the oil had to be transported to market by horse-drawn carriages or rail, inviting the risk of spills. But ultimately steel solved this problem, too, with pipelines creating a safer way for the oil to get from the source to the customer.[*]

Pittsburgh emerged as a globally important city, a place where coal, steel, and oil all came together. One of the most famous steel barons was Andrew Carnegie, whose philanthropy in support of libraries, museums, and universities makes his name still widely known today. His company ultimately became US Steel in a buyout organized by J. P. Morgan. US Steel is headquartered in Pittsburgh today and was the world's first billion-dollar company. Reflecting the importance of steel, the city's National Football League team's name is the Steelers. Ultimately, steel enabled boilers, pressure vessels, and internal combustion engines, new devices that could convert heat into motion or other useful outcomes. Materials such as wood would burn up or fall apart from the high pressure and temperatures of combustion, making them unsuitable for high-performance machines. The rise of the internal combustion engines made from stronger metals enabled mass transportation through personal vehicles. Consequently, the demand for oil increased even further. Coal and steel unleashed a wave of ultra-wealthy entrepreneurs, but oil, which could be produced with smaller operations and in much more widespread fashion, created even more wealth and helped foster the rise of a larger middle class.

## THE TRANSISTOR, THE MICROPROCESSOR, AND THE INFORMATION AGE

Sharper and better tools from coal transformed society. Oil and gas gave us plastic and other materials that changed society

---

[*] Relatively speaking, pipelines fail rarely, but when they fail, the spill can be significant. Rail cars spill more often, but their spills tend to be smaller. Thus, neither transportation mode for oil is perfect, but comparatively speaking, pipelines are safer and lower risk overall.

once again. But the rise of electricity gave rise to even more so-phisticated tools for manufacturing and information storage and processing that seemed to accelerate the pace of change. The high-tech industry cannot exist without electricity; it's a mod-ern sector and therefore needs a modern form of energy. Those parts of the world that lack access to modern forms of energy also do not have those modern sectors, because you can't run an IT sector or its appliances directly on fuelwood, for example.

At the core of the information age is the transistor. This lit-tle device was invented at Bell Labs in 1947. At its simplest, the transistor is just an electrical switch that is either on or off. This state could be used to approximate a binary number, with 1 equal to *on* and 0 equal to *off*. A series of transistors could cre-ate a series of binary digits, or bits, which if combined the right way form the underpinning of integrated circuits (electrical circuits with multiple components like resistors and capacitors embedded within a chip) and microprocessors. The micropro-cessor could automate all sorts of functions. Initially, it was used for complicated math problems that were too time-consuming and cumbersome to solve by hand. Today, microprocessors have taken on an incredible number of other computational func-tions, such as displaying particular colors on a computer mon-itor or controlling equipment. At its launch it was an exotic technology and laboratory curiosity, but today it is embedded in our appliances and consumer electronics. Smartphones, modern cars, and video game consoles all require microprocessors to op-erate, and now even passive appliances such as light bulbs might include small microprocessors to enable remote operation, sens-ing, or adjustability.

These tools are now so ubiquitous that it is hard to imagine life without the electricity-enabled information accessible at our fingertips. The information technology boom, like the coal, steel, or oil booms before it, also created wealth. The modern version includes tech billionaires—Bill Gates, Bill Hewlett and David Packard, Steve Jobs, and others—who made their money from computers powered by microprocessors running on electricity. The electrically driven transistor underlies this modern wave of wealth creation, just as oil-driven combustion engines or coal-driven factories created prior booms. US Steel was the world's first billion-dollar company, but Apple was the world's first trillion-dollar company.

The cumulative energy needs of the IT industry globally are quite large. About 2 percent of national electricity consumption is needed to power racks of computers in data centers and the air conditioners needed to cool the computers.[6] One rack at a data center, which looks like an open filing cabinet filled top to bottom with high-speed computers, consumes the same amount of power as a neighborhood of homes. A data center or server farm has hundreds of these racks all lined up in neat rows. Those microprocessors generate a lot of heat when they operate, so a lot more electricity is used just for cooling them back down.

The information-energy nexus is interesting and follows paradoxical pathways. As information becomes more energy-intensive (because of the data centers), energy becomes more information-intensive (because of the additional data available from smart meters, sensors, and other devices that let us keep close tabs on the energy sector). And, ironically, information is going from distributed (on our desktops) to centralized (via the cloud), while electricity systems are doing the opposite, going from large centralized coal plants to smaller, distributed rooftop solar arrays or backup generators.

That electricity enabled an information economy did not just create wealth for those entrepreneurs, but also made the markets operate more efficiently, which saved money economy-wide. So the cascading effects are significant.

In addition, electricity opens up the possibility of new kinds of materials and new kinds of tools. For example, modern electricity can be used to produce graphene—a sheet of graphite one atom thick—which opens a pathway to new materials that are stronger and lighter than ever before. These materials can be used to make supercapacitors or other storage devices, lighter devices that don't require as much energy when transported, and more efficient electric components. Electricity also is used to make lasers for precision welding, cutting, and ablation (for instance, to smooth metal or conduct surgery). Old steel mills that used heat from coal are being replaced by modern mini-mills that use electric arc heaters to smelt steel, saving energy and achieving higher performance.

Critically, electricity is necessary for 3D printers, which are changing the world of manufacturing. Machine tools are considered the mothers of machines because they are used to make everything else. Lathes, drill presses, and mills allow gifted machinists to turn blocks of metals into sophisticated components. The aerospace and automotive industries wouldn't be possible without machining. But those machines are subtractive—just as a sculptor starts with a big block of marble and then carves it down to the desired shape, machinists start with a block of metal and remove the parts they won't want by drilling, cutting, shaving, and milling. The revolution of 3D printers is that they are additive. Instead of starting with something big and subtracting excess to make it smaller, additive manufacturing starts with powders and builds them up into the desired 3D object. Additive manufacturing can build pieces impossible to make with machining: it can build a castle with spiral staircases inside it, something that would be impossible the traditional way. While 3D printers require a lot of electricity, because they are small they could be distributed closer to the customer, avoiding transportation costs for delivering products. Instead, the 3D printers will be so affordable, people will just print their products at home. Fossil fuels took us from an era of custom products made by master craftspeople toward mass production with higher quality and more consistency from product to product. But 3D printers combine them to offer mass customization instead of mass production. They are just one more reminder that energy, innovation, and manufacturing are all intimately related and in the process create wealth and jobs.

## ENERGY JOBS

As the energy sector develops and becomes a larger part of the world economy, it also creates sought-after jobs that tend to pay better than other manual labor opportunities in factories or on farms. The energy sector has been a major employer since its inception, and that makes it economically desirable to planners and political leaders. It is also a capital-intensive sector where many of the struggles for workers' rights played out, making the energy sector a battlefield in the tension between labor and capital. Powerful unions are associated with coal mining, electrical work, and oil and gas production. Interestingly, Soviet dictator Joseph Stalin got his start as a union leader in the oilfields of Baku.

Today the energy sector employs more than 20 million people around the world. Nearly 10 million are employed by the renewable energy industry alone, and another 2 million are employed just for oil and gas in the United States.[7] Those workers include those who are working directly at wind farms, coal mines, or drilling sites for oil and gas; those employed at retail fuel stations; manufacturers of specialized machinery; construction; and related support jobs like lawyers, bankers, and researchers. The ability of the energy sector to put people to work while making resources available that create wealth is an attractive combination. It is also why it is not unusual for politicians to devise major energy development projects to stimulate the economy.

During the Great Depression, Presidents Herbert Hoover and Franklin Delano Roosevelt wanted to boost employment. Roosevelt looked to the Works Progress Administration (WPA) as a way to invest in job creation through public works, including building offices and trails at state parks and many other types of public infrastructure that still stand. It also included building massive dams that would serve multiple goals at once. A dam could enable irrigation, recreation, navigation, and power generation. Killer floods in the eastern part of the United States made planners eager to tame wild rivers to reduce the risk of death and destruction. At the same time, the US Army Corps of Engineers and others wanted to facilitate more economic growth and commerce by using inland waterways to move cargo on barges. In the semi-arid West, farmers with fertile soils desperately wanted irrigation for their thirsty crops. Power generation was often an afterthought at these dams, with the expectation that sales of electricity could be used to cover the costs of the dam itself. Today hydroelectric power is the biggest source of renewable electricity globally.

These dams employed many people during construction, offering an economic boost. The Hoover Dam, initiated during the Hoover Administration in 1931 and completed in 1935, employed over 20,000 people just for construction. The water for irrigation gave another economic boost from crop production. Other dams facilitated easier trade because they incorporated a series of locks and canals that offered another economic boost by lowering the cost of transportation. Then the access to cheap electricity was a final boost yet again. It is for all these reasons that politicians find hydroelectric dams such appealing development efforts: their construction provides immediate economic gain and then their ancillary benefits add an ongoing economic stimulus.

In 1920s Ireland, a similar phenomenon was afoot with the creation of a large hydroelectric project on the River Shannon that helped electrify the country. Though at 85 megawatts the power plant is not that significant by modern standards, at the time the project required 20 percent of Ireland's national revenue, a massive undertaking.[8] It employed thousands of workers and triggered the creation of the world's first national electricity network. The Shannon Scheme was partly justified because of its economic impact. An ad produced in 1927 to generate public awareness and popular support for the project noted that the project would provide 90,000 horsepower. It also complained that

> The American workman is the most prosperous on earth, because he has, on an average, three horse-power, the equivalent of thirty human slaves, helping him to produce.
>
> No wonder he can toil less and be paid more than the workman of other lands. He is not a toiler, he is a director of machinery. . . .
>
> Shannon electricity will lift the heavy work from human shoulders to the iron shoulders of machines.[9]

For the Irish, the hydroelectric dam offered a way to catch up with the American worker, who had high productivity because of the electric motors that assisted him. Similar stories played out in other countries looking to modernize their economies while employing people and pursuing wealth creation. In Southeast Asia, major dams have left their mark. One of them shifted the direction of the Tonlé Sap, a major river famous for running backward part of the year. In Egypt, construction of the Aswan High Dam on the Nile River required relocation of ancient monuments and affected the silt distribution patterns of the river. Another dam project is currently under consideration on the Congo River, the second-largest river by volume in the world. Its proposed scale of 40 gigawatts at full build-out is twice as large as China's world-tilting Three Gorges Dam. The Grand Inga Dam is controversial because—as with the great American or Chinese dams—it would displace tens of thousands of people and forever leave its scar on the impacted ecosystem. But developers hope it would also employ many people and help electrify the African continent, generating many follow-on economic benefits. It is not clear whether the benefits outweigh the costs, but because of a long legacy of corruption there is concern that the benefits will not flow to the people anyway, so the project remains in limbo.

Nearly seventy-five years after the Hoover Dam's creation, while the United States was reeling from the Great Recession that started in 2008, energy was again sought as a tool for economic resuscitation and employment. President Barack Obama pushed for an economic stimulus package that included tens of billions of dollars for investment in energy projects to kick-start employment in a way that would have lasting benefit in subsequent decades. Those investments targeted modern renewables such as wind and solar power, installation of smart meters for home energy monitoring, battery manufacturing for electric vehicles, home weatherization, nuclear power, and clean coal.

Like other sectors, energy is a great job creator. But it also can destroy jobs in other sectors and suffers its own waves of job losses on occasion. The rise of modern energy and its companion machines reduced the numbers of farmers needed to feed society, of lumberjacks needed to produce timber, and of trainers needed to break horses. Energy's successes also undercut the energy sector's employment. The coal-mining sector employed

863,000 people at its peak in 1923, after which it endured a century-long employment decline. There was a boost in coal-mining employment in the 1970s after the oil crises made power plant operators wary of using oil and federal policy outlawed the use of natural gas for new power plants. As a result, more than 80 gigawatts of coal plants were built that otherwise might not have been, creating a surge in demand for coal, with a matching surge in employment. But the sector continued to get more production per employee by going to larger mines with bigger equipment, and the decline in employment resumed shortly thereafter. Surface-mining approaches such as mountaintop-removal mining leave an incredible scar, especially in hilly areas like the Appalachian Mountains of Pennsylvania, Kentucky, and West Virginia, where coal mining is prevalent. But they also use massive machines to scoop off the mountain's surface to get to the coal seams beneath. One machine does the work of many miners.[10] Now with smarter algorithms and data analysis, the machines are even more productive. Thus the rise of machines and microprocessors enabled by modern electricity powered by coal becomes the coal sector's own undoing.

## HEALTH

In some respects, health is the ultimate expression of wealth. The point of having a wealthy society, in addition to its freedoms, is to live a long and healthy life.

Energy enables better health care in a number of ways. One is the use of energy to heat water for basic cleaning and hygiene. It is not unusual for hospitals to have energy-intensive steam systems for sterilization and cleaning of laundry or medical tools. That hygiene reduces the risks of infection. In 1761, a hot bath in Paris cost about 3 pounds. But the daily wages for a craftsman were half a pound a day. So they would have to work for six days just to earn enough money to take a bath.[11] Access to modern energy and water systems lowered the cost of personal hygiene to the point at which today it is far down the list of monthly expenses despite a higher frequency of bathing.

Lack of hygiene and sterilization could be deadly. As depicted with grisly detail in *The Butchering Art*, a book by Lindsey Fitzharris about the state-of-the-science for surgery in Victorian England, hospitals were deathtraps. Patients might survive the surgery but would frequently die from infection during recovery. It wasn't until a kind Quaker doctor named Joseph Lister slowly and methodically revealed to the world the importance of germs and disinfection that success rates improved. Because of his breakthrough, the production of antiseptics became a booming industry, with one enterprising company even naming its product—*Listerine*—after the doctor.

Lister used phenol—then known as carbolic acid—as his antiseptic of choice. Phenol was originally made from coal tar but today is made on a very large scale (nearly 10 billion kilograms per year) from petroleum. Phenol is now a precursor for many pharmaceutical drugs, throat sprays like Chloraseptic, and lip balms like Carmex. In this way, fossil fuels improved the delivery of health care and medical outcomes.

In some ways using petroleum to make modern pharmaceutical products is in keeping with an age-old tradition. Oil was used as a medicine for animals and humans for millennia. Ointments for soothing sore skin have been used for a long time. The initial production of oil in Titusville, Pennsylvania, in the 1850s was to get rock oil for medicinal purposes. Standard Oil's ads in the early 1900s made bold claims about the medicinal value of petroleum products. And "petroleum jelly" and other similarly named personal care products hint at the petrochemical roots of medicines. It continues today when modern pharmaceuticals are made from chemicals produced from natural gas and oil.

Surgery enjoyed other improvements from the availability of modern energy forms. Surgical rooms were poorly ventilated and dimly lit in the Victorian era. Performing surgeries by candlelight was a common occurrence, which was problematic because it was dim and flickering, casting shadows that moved, no doubt complicating the precision of the procedure. It was not unusual for patients to suffer the indignity of hot wax dripping on them after the doctor would bring the candle dangerously close so that the patient could be inspected. Other options for illumination were natural lighting through skylights—which could only work at certain times of day and if the weather cooperated—or from fireplaces and oil lamps, which along with candles produced fumes that contaminated the room's sterility and could compromise the health of the patient undergoing surgery.[12] The medical profession was well served by the invention of fumeless, bright, steady electric lights and fans for ventilation.

Modern energy also produced better tools, including sharper scalpels made from steel and plastic devices that were inert and easier to clean and keep clean. Ultimately, even better diagnostic tools such as X-rays, magnetic resonance imaging (MRI) machines, ultrasound, and nuclear medicine were enabled by electricity, giving doctors the ability to see through the skin and into the body. New surgical tools, such as lasers for eye surgery or robotic assistance for surgery (to minimize the size of incisions), have also improved the precision with which surgical work can be executed.

The extent to which our modern medical treatments depend heavily on the availability of electricity was revealed after hurricanes ripped through Puerto Rico in 2017, knocking out the power and gutting the island's capability to offer high-quality medical care. To minimize these risks, many hospitals have their own backup power systems on site.

## LIBERATING WOMEN

Of particular importance to the economy was the way electrical appliances liberated women from some of the toil of household chores such that they could participate in the private workforce. In many ways, the second wave of women's liberation in the mid-1900s went hand in hand with the rise of electrical appliances such as mixers, vacuum cleaners, dishwashers, and clothes washers, which made domestic household chores much less time-consuming. As a result, women had more opportunity—and perhaps more desire, too—to be part of the workforce. It just might be that access to modern forms of energy—in this case, electricity—and new appliances helped the cause of women's rights more than is typically understood.

So what does that mean for women around the world? There are many millions of women in sub-Saharan Africa and Southeast Asia who toil by hand for hours to tend to farms and fetch water or firewood, and in the process miss out on the opportunity to get an education, seek jobs, or launch their own businesses away from the farm. Just as modern forms of energy improved women's freedom in the United States decades ago, the same would be true for those women around the world. That makes energy not just a concern for the environment or national security, but for women's rights, too.

The idea that modern electrical appliances were liberating for women shows up in our popular culture. Loretta Lynn, a famous country singer and songwriter who broke barriers with her work, wrote the song "Coal Miner's Daughter," which was also the title of her autobiography and a movie starring Sissy Spacek. In theaters in 1980, this movie shows the struggle of a dying coal community from decades earlier when there were few economic options for men, and even fewer for women.

This theme is familiar in modern coal towns, as it is a pattern that has been perpetuating itself since coal mine employment peaked in the United States in the 1920s.

In one notable scene, Loretta Lynn acquired a washing machine to do the laundry. While the machine washed the clothes, she would sit on the porch and practice her music. This new electrical appliance gave her time to write and practice her songs despite being a busy young mother. In the animated children's movie *Sing*, which came out in 2016, the same type of scene plays out. A mother pig, voiced by Reese Witherspoon, creates an automated machine to serve food and wash dishes for a full house of piglets, giving her time to pursue her dreams of a singing career. Why would both of these movies make the same visual reference to the idea that electric appliances would liberate women? Because it's an idea based in reality.

Even the River Shannon hydroelectric project introduced earlier considered women to be primary beneficiaries of and stakeholders in the project. The Electricity Supply Board released a series of ads targeting women in the 1930s to generate support for the project—a de facto admission about their important role as key decision-makers in the home.[13] One ad's boldface headline blares "A Valiant Woman," and the text goes on to explain the importance of electricity as modern help for modern women and then concludes with a call for women to get their homes wired for electricity. This point is hammered home with a partner ad that declares electricity lets women operate the houses with greater speed, giving them more leisure time.

A documentary titled *Hydro*, produced by the Bonneville Power Administration in 1939 to promote the development of dams along the Columbia River, showed the various ways in which women would benefit from electric appliances in the home, extolling the virtues of "an electric home with more leisure and better living."[14] Sure enough, a subsequent ad shows a happy, well-dressed woman casually reading a book illuminated by an electric light. The ad goes on to note that because of electrical appliances, we can see "the Mistress of the Home free from unnecessary toil—free from health-destroying fatigue—free to rest and read and recapture that sense of buoyant joy in life." There is no hinting with this ad: electrical appliances save women effort and free them up for other activities. One ad for Christmas 1930 called for electrical gifts, showing a well-dressed woman with a Christmas package and a list of gift suggestions for her and her friends, most of which were electrical appliances that women would presumably find desirable, such as an Electric Fire, Electric Iron, Electric Toaster, and Electric Tea Kettle.

## ENERGY PHILANTHROPY, DEMOCRACY, AND GLOBALIZATION

Energy access is a democratizing force because it opens up prosperity to more people rather than concentrating it in the hands of the aristocratic few. The democratization of wealth didn't happen everywhere: countries like Iran, Saudi Arabia, and Russia have significant wealth concentration. The stark difference between North and South Korea illustrates this idea further. Images from space show Southeast Asian cities lit up at night because people have access to electric lighting. However, by contrast, North Korea is dark except for Pyongyang, the capital city. Where people have access to energy, they have more freedoms than in places where they do not have access to energy.

In the United States, energy consumption and the middle class grew together in the twentieth century, partly because the wealth from energy production became available to many more landowners and energy company employees. In other countries, the story is different. For example, countries like Russia and Saudi Arabia that have nationalized energy companies rely heavily on energy sales to fund their budgets. The United States, by contrast, mostly relies on tax revenues. Just as the idea of "no taxation without representation" was a motivating call to action in the Revolutionary Era of the late 1700s, the opposite—"no representation without taxation" is relevant today. What that means is in the United States, because the government relies on citizens' taxes to function, the government is ultimately accountable to voters. But in those countries that rely on oil revenues instead of taxes, the governments are less accountable to their citizens. That introduces the risk of oppressed freedoms. It also puts those countries at risk of economic failure if energy prices fall. In the aftermath of the 2015 oil price collapse, Russia, Venezuela, and Middle Eastern oil-producing nations all struggled to pay their bills. A tax-dependent nation like the United States relies less on the price of oil, but energy prices still matter. When they are low, they are an economic accelerant nationally because they encourage consumption and production in energy-intensive industries. But when energy prices are high, they encourage capital investment in the energy sector. So both high and low energy prices are good for a country like the United States.

One of the benefits of sudden and dramatic wealth generation was that the energy industry created a fleet of philanthropists. Many of the biggest names in American philanthropy in the twentieth century had their roots in energy industries or industries that depended on energy. Carnegie made his money in steel, whose fate was intertwined with coal. Today, major universities, libraries, and institutes still bear his name.

Howard Hughes, who was born rich because of his dad's drill bit company (known as Baker Hughes today), became even richer with his stewardship of Trans World Airlines, an aerospace company, and movie production. At one point he was the richest man in the world. He became quite eccentric in his old age but also became a major philanthropist, establishing the nonprofit Howard Hughes Medical Institute. HHMI's endowment was $18 billion in 2017, making it the second-largest philanthropic organization in the nation. It gives away more than $1 billion annually for medical research and other causes.

Other prizes and names are also quite familiar. The Getty family, whose company Getty Oil made them very wealthy, built the landmark Getty Museum in Los Angeles, with a crowning view of the basin. Even the Nobel family, whose endowment awards the prizes recognizing great scientific achievement, made their money from dynamite and other equipment for the oil industry.

But the most important philanthropist was John D. Rockefeller. He, and especially his son, John D. Rockefeller Jr., set philanthropy on a new course by making their giving a systematic process. Ron Chernow's book *Titan* reveals that Rockefeller always gave charitably. Even when he was very poor, he set aside 10 percent of his income for charity based on his Baptist upbringing. As he got richer from selling kerosene from crude oil, his giving scaled up. When he got richer again from selling gasoline for the internal combustion engine, his wealth and his philanthropy grew further. At one point he was the world's wealthiest man, and his philanthropy was legendary. But compared to other notable philanthropists of the era, Rockefeller did not like to put his name on buildings or institutions. Those places that have his name were the handiwork of his son, as the patriarch considered charity to be an expression of humility rather than something that should be crowed about. If only all philanthropists followed that pattern.

One of the reasons Rockefeller became so wealthy is that the oil industry tapped into a global market. Energy is unique in that it is both a global market—crude oil is traded in every country, for example—and it is the enabler of a global market. Energy is used to move goods from one place to another, and is often one of those goods that is moved. Global trade is good news in many respects—bringing economic efficiency and benefit to many corners of the world—but it also has some downsides.

Because the modern economy—and so much wealth creation—depends on access to affordable, reliable energy, energy shortages or price shocks can be very destructive to the economy. The latter can take many forms. In 2008, oil prices spiked, triggering a global economic recession. In the United States, after gasoline prices nearly doubled, some consumers had to choose between paying for gasoline and making their mortgage payments. Most chose gasoline because they needed to drive to work, and delayed or skipped their mortgage payments, helping to trigger the mortgage and home loan crisis.

In the 1980s, the oil price shock was the other way around. After Saudi Arabia flooded the world oil market with crude, oil prices collapsed. In oil-producing states like Texas, the collapse in oil prices bankrupted many producers and led to mass layoffs. Because many loans were backed by oil assets and those assets lost value when the oil prices collapsed, banks started to collapse. Tens of thousands of smaller banks called savings and loans (S&Ls) failed, creating a domino effect nationwide. Eventually, lending laws were rewritten to toughen the standards to help avoid similar consequences from future price drops. That the oil price drop that started in late 2014 did not have a similar effect is a sign that those rules might have been effective.

Overall, price spikes up *or* down can be problematic, and that economic risk opens up consumer nations to the vulnerability that producing nations will use oil price as a weapon. (See Chapter 6, "Security" for more on the oil weapon.)

Given all these economic impacts of energy, the question for humanity is how to get everyone the benefits of a wealthy existence—comfortable homes, freedom of movement, access to information, liberty from manual labor—without consuming so much energy and causing so much environmental impact. There are a few basic approaches. First, find more efficient ways to achieve the same task or energy service. More efficient air conditioners or refrigerators, windows that insulate better, and cars with better fuel economy can preserve food, maintain home temperature, and move people and goods with lower energy requirements. Second, promote behavioral approaches that use devices less: turning off lights when leaving a room and taking mass transit to work, for example. Third, use cleaner forms of energy. After efficiency and conservation have reduced how much energy is required, if energy is used in a cleaner way, the environmental impacts can be reduced. This might be achieved by adding scrubbers to smokestacks and tailpipes to remove pollutants before they are emitted into the atmosphere. Or it can be done by using alternative forms of energy that are cleaner.

These approaches are critical, because even though access to energy can make us richer, using too much energy that pollutes can make us poorer. At some point if the air and water are very polluted, the economic toll becomes expensive. Millions of people die prematurely from air pollution in China. Many more have a reduced ability to work because they are afflicted with pollution-induced sickness such as asthma, or they must tend to loved ones who are sick. Those deaths and sick days hinder the economy. Ecosystem damage from pollution reduces the money that can be made from tourism, hunting, fishing, and the timber industry. While energy access is clearly a pathway to wealth for someone in energy poverty, for societies that are already rich, unfettered access to and use of dirty forms of energy make it harder for the economy to grow.

Chapter 5

# CITIES

The world's population has doubled in the past fifty years, from 3.5 billion to over 7 billion people. In parallel, the global economy has grown. As the population increases, more people are abandoning rural areas for cities to pursue a different quality of life. This trend has continued for almost a century; today less than 20 percent of the US population resides in rural areas. China has about forty-five cities with over a million residents, and that number continues to increase.

These cities are all built on energy. Without modern energy forms, the concept of a modern city would be impossible to fathom. Energy controls the lighting, temperature, and humidity of those indoor spaces, making them comfortable. Energy is necessary to construct the buildings that accommodate high population density by extending toward the sky. And electricity is critical for the lighting, the elevators, and the water pumps that make living indoors in high rises possible and comfortable.

At the same time, cities rely on transportation to bring goods and food in and take wastes out. It is no surprise that the rise of the megacity coincided with the industrial revolution, which coincided with a rise in newer, energy-dense fuels such as coal and petroleum. The story of modern cities and modern energy go hand in hand. As our cities evolve, our energy systems will evolve alongside. Whether energy improvements enable new cities or new cities enable new forms of energy isn't clear, but their close partnership is undeniable.

Cities are inherently a densification of population, which requires the active management of water and the concentration of energy supplies. We wouldn't have cities in societies comprising hunters and gatherers, whose energy in the form of food and firewood was dispersed. Once the ability to grow sufficient food in agricultural locations emerged, water supplies were firmed up, and concentrated energy was available, it became possible to support more people in one place, and cities could thrive.

David Sedlak noted in his book *Water 4.0* that there have been several eras of water management in human history.[1] Stable water management led to increased food production without wandering, which meant civilizations could stay put and then specialize. The rise of sanitation systems and water treatment allowed dense cities to form. If water enabled city densification, electricity made cities safe (with public lighting) and allowed cities to move upward toward the sky.

Many people continue to migrate to cities worldwide, which puts cities in a prime position to solve global resource problems. Mayors are taking on more responsibility for designing solutions simply because they have to, especially in countries where national enthusiasm for tackling environmental issues has cooled off.[2] The international climate agreement forged in Paris in December 2015 also acknowledged a central role for cities.[3] More than 1,000 mayors flocked to the French capital during the talks to share their pledges to reduce emissions. Changing building codes and investing in energy efficiency are just two starting points that many city leaders said they could initiate much more quickly than national governments.

It makes sense for cities to step up. Some of them—New York, Mexico City, Beijing—house more people than entire countries. And in urban landscapes the challenges of managing our lives come crashing together in concentrated form. Cities can lead because they can quickly scale up solutions and because they are living labs for figuring out how to improve quality of life without using up the earth's resources, polluting its air and water, and harming human health in the process.

## INDOOR LIGHTING, SKYSCRAPERS, AND CLIMATE CONTROL

The key ways in which energy improved indoor life were lighting and climate control. The key way in which it enabled urban living was by allowing for densification through transportation networks and taller structures, both of which require energy. Because of a human's limited ability to see in the dark, indoor lighting had been desired for millennia. Torches, fires, and candles had been used, slowly giving way over time to slightly more advanced forms of lighting. In rich homes, such as the palace at Versailles, light was used as a display of wealth. For example, over 24,000 candles were used in 1688 to light a park at Versailles.[4] They used beeswax for candles, even though it was very expensive, because it would not generate smoke.[5] Oil lamps— sometimes using olive oil in the Mediterranean or whale oil in other locations—were brighter than candles. Lamps and lanterns were developed using a variety of innovations, including hollow wicks, glass containers to protect the flames from wind, or lenses or other optical elements that would project and cast the light farther out.

Because lighting was a desired service but difficult to procure, it was often fantastically expensive. Human Progress, a project set up by the libertarian think tank Cato Institute, noted that "the amount of labor that once bought 54 minutes of light now buys 52 years of light." And that "George Washington calculated that five hours of reading per night cost him over $1000 in today's dollars."[6] The introduction of fossil fuels in the 1800s and the invention of newer, better forms of lighting—especially gas lighting and then the electric light bulb—changed everything again. Energy brought us light, liberating humans from the shackles of darkness.

Whale oil was a particularly popular illuminant because it burned brighter and cleaner than wood. However, it also produced a pungent odor, and after the mid-1800s, whaling became more expensive as whale populations were depleted by the global desire for whale oil. Around that time, crude oil production was revolutionizing lighting. A barrel of crude oil can be converted to many useful products, including gasoline, diesel, wax, and tar. It also produces kerosene, which is an even better illuminant than whale oil because it burns even brighter and without the odor. And because whales were far away and expensive to harvest, kerosene was a closer, cheaper, better alternative for oil lamps. Noted energy historian Daniel Yergin called Rockefeller, who became the world's richest man from sales of kerosene, a "merchant of light."[7]

In parallel, for highly urbanized areas like London and Paris, natural gas distribution systems were built. Homes could connect to the pipes to have natural gas indoor lighting, heating, and cooking. A key at the wall could be turned to let the gas flow through the lamp, and a sparker used to ignite the flame. While these gas lamps were brighter and cleaner, they still produced fumes and sometimes blackened interiors with smoke and soot. They also had a bad habit of triggering fires on occasion.

Electric lighting was a major advance because it brought higher-quality light, without the flicker, while avoiding the fumes and reducing the risk of fire. An ad from the 1930s associated with the River Shannon project extolled the virtues of electric lighting as a safe alternative to candles or gas lamps, noting that even children can operate the light switch without risk of being burned.[8]

Once people moved closer together, they began to rise higher. Chicago was the first modern city with a profile graced by skyscrapers. Those skyscrapers could not have existed without modern energy. Coal was critical to making the steel that allowed builders to liberate themselves from stone building materials and consequently reach new heights. Before that, the stone walls themselves held up the load of the building. That meant taller buildings needed very thick walls and were limited to about a half-dozen stories. But with the advent of steelwork skeletons using I-beams to bear the load instead of the walls themselves, buildings could reach higher.

One problem with taller buildings is how to get people, goods, and water to the upper floors. It is hard to imagine lifting or carrying all of that water by hand to a twentieth-floor apartment. That's where electricity comes in. Not only did it provide the lighting that made indoor living more desirable, but it also made it possible to pump water to rooftop tanks. Once the water was in the tank, it could flow by gravity downward to the various taps and faucets on the different floors.

Indoor life also needed climate control. Creating shelter is nothing new. But for those populations who lived in colder climates, keeping alive through harsh winters was a nontrivial task. Human beings' control of fire is one thing that distinguishes us from other species. And that fire was critical not only for cooking, as noted earlier, but also for warmth. Primitive fuels like wood, straw, cow dung, and peat generated heat from simple configurations like open firepits, but also a lot of smoke, soot, and ash. Hence, there was a need for higher-quality fuels and heating systems as we moved into more tightly enclosed spaces.

Fireplaces that burned wood or coal and were lined with bricks at their back helped radiate heat into the room and used chimneys to whisk smoke away. But there was still room for improvement. Benjamin Franklin's invention—called the Franklin stove—used fuel more efficiently while providing indoor heating and a surface for cooking. Eventually, the Franklin stove was replaced by more modern gas, fuel oil, or electric home heaters. Ironically, some environmentally conscious consumers are moving back to fireplaces or heaters fueled by wood pellets to avoid the use of fossil fuels.

This is all well and good for cold weather, but what about when it's hot? Air conditioners were once a luxurious modern convenience for living indoors, especially in urban dwellings with fewer breezes or in hot and humid climates. Their current ubiquity contributes significantly to energy consumption. Today, the electricity consumed in the United States for air conditioning could meet about half the total electricity demand for all of Africa.[9] Climate control made different regions and ways of living tolerable. Air conditioning was also important for improving working conditions in factories.

Heating a room can be accomplished simply by burning things, but cooling a room required a sophisticated understanding of thermodynamics to come up with a vapor compression refrigeration cycle that could use a clever combination of liquid refrigerant, compressors, and heat exchangers to move heat from inside a house to the outside. It is the same process that moves heat from inside a refrigerator into the surrounding room. The inventor of the modern air conditioner is a man named Willis Carrier.[10] He invented the modern electric air conditioner (with humidity control) in 1902, and the namesake company he founded in 1915 went on to manufacture them. Air conditioning saves lives, increases the livability of hot regions, and enables better-functioning schools and hospitals.[11] The entertainment industry helped spur this innovation, as air conditioning made windowless movie houses much more comfortable for moviegoers who appreciated the opportunity to escape the summer heat. Movie theaters were a typical way for people to get their first exposure to air conditioning, helping to drive its popularity.

Where available, air conditioning relieves suffering. And because of climate change, hundreds of millions of people in hot, humid climates are projected to endure worsening heat that will put lives at risk. As a consequence, the demand for air conditioning is on the rise globally. In India it shows up as the proliferation of window units that bring comfort to cramped, stuffy apartments. In the United States, the air conditioner has evolved from a desirable luxury item to practically a necessity in consumers' minds.

The air conditioner not only improved our comfort indoors, but also triggered the great southward migration to the Sunbelt in the United States. The region had languished because of its uncomfortably hot and humid summers, but with air conditioning providing a cool escape from the outdoor climate, the South looked more attractive, and its population has grown consistently since the end of World War II.

Air conditioning is now so common that it is the single biggest driver of peak electricity demand in hot climates. In a place like Texas, almost half of the power consumption statewide on a hot summer afternoon comes from air conditioners in homes, which strains the grid and raises the risk of blackouts. That makes air conditioners a driver of global warming, which of course raises the need for air conditioning. And the refrigerants used inside the air conditioner to remove heat from homes are also greenhouse gases, so as air conditioners proliferate, so does the risk of leaks, which can worsen the abundance of heat-trapping gases in the atmosphere.

## STREET LIGHTING

Indoor lighting made homes more enjoyable, opened up educational opportunities, and enabled factories to work more shifts, but street lighting was also critical to city life. Streetlights made it easier for city inhabitants to stay out late, creating the concept of "nightlife," which had not existed before, and making cities safer.[12] In many ways, lighting and public safety were necessary preconditions to modern urban life.

Before public lighting, people had to carry their own lights to go out at night. It was not unusual for con men carrying torches to offer wayward strangers or visitors their services for a fee to light their path through a dark, unknown city after sunset, only to lead them into an alley where their colleagues would rob them.[13] The rise of city lighting reduced those risks and made the streets safer. That meant restaurants could stay open later, and stores could, too. Stores went a step further, lighting up their window displays to attract customers the way a light attracts moths. P. T. Barnum learned this lesson well, and used a lamp from a lighthouse to attract audiences to his circus.[14] Perhaps that experience gave him an appreciation for lighting. When he was mayor of Bridgeport, Connecticut, he pushed to bring gas lighting to the streets.[15]

Public lighting took many forms, including gas lamps and then electric lamps. Gas lamps were distributed around the city and connected to underground gas mains starting in the early 1800s in London.[16] Because lighting the gas lamps was a dangerous task and could lead to an explosion if done incorrectly, a fleet of professionals—lamplighters—had the job each night of safely lighting the lamps to illuminate the cities. To this day, there are still five of them who have that task in London.[17] However, most of the world's gas lamps have been replaced by electric lighting systems that are brighter, cheaper, and less likely to catch fire.

In Austin, Texas, city lighting was initially provided by iconic moontowers. About two dozen of these 150-foot-tall towers were placed around the city, and seventeen of them are still standing today. The city installed them to cast light over a larger radius, avoiding the need for a lamp at every corner. Other cities that installed moontowers have moved on to more precise lighting approaches, and now the remaining moontowers in Austin are historical icons because of their status as oddities. The Richard Linklater movie *Dazed and Confused*, which was actor Matthew McConaughey's breakout film, features a classic party scene under one of these moontowers.

Though street lighting transformed cities in a way that was liberating and safe, it also had its downside, namely light pollution. The ease and low cost of lighting means it grows in application, exacerbating the problem. Dark skies are a lost phenomenon for urban dwellers worldwide, making it harder for people to see the Milky Way. When I lived in Los Angeles, which is brightly lit and cloaked in a thin layer of air pollution, even the brightest constellations were hard to see. The running joke for locals is that the L.A. Star is actually just the moon, the only celestial object visible from the ground.

My wife and I sit down with our children on New Year's Day each year to set goals and resolutions for the year. Among our many personal and professional goals, we also include family goals—for example, to travel somewhere new together. One of our annual goals is to see the Milky Way. That such a goal takes effort is a sign of its difficulty. We have to intentionally plan to go to the countryside for long enough that one of the nights will have clear weather; otherwise we won't see it. The loss of dark skies is not just an aesthetic problem for my family: it also disturbs reproduction in coral reefs or animal species such as sea turtles.[18] It affects bird migration when millions of birds die from collisions with illuminated buildings and objects.[19] In response, environmentalists have sought to point the lights downward, reduce their use, and minimize their glare to reduce their impact on the skies. The color of lights can also be adjusted to prevent wildlife impacts: green lights on offshore oil platforms or other industrial equipment help prevent disturbances to bird migration, blue lights on planes can reduce the incidence of bird strikes, and color contrasts on windows lower the rate of collisions.[20]

Though street lighting has been around for centuries, it is still an evolving technology. Cities are now replacing traditional light bulbs with newer light-emitting diode (LED) versions that are much more efficient and last much longer. That means they will reduce energy consumption, which is good news, and they will save money because they will need to be replaced every couple of decades instead of every couple of years. Using LEDs can lead to layoffs, unfortunately, as bulb-changer laborers are less necessary.

But lights have another unique feature: they are distributed around urban areas and tend to be concentrated where the people, traffic, air pollution, and crimes might occur. Smart lighting—that is, equipping lights with data collection capabilities—presents an opportunity to get a lot more information about a city, improving public health and safety. Placing air quality sensors on the streetlights helps give real-time and finely resolved information about air pollution that can be used for mitigation programs like ozone-action days and incentives for people to use mass transit or carpool. Acoustic sensors are also a new feature. Microphones can pick up sudden noises such as gunshots to help first responders dispatch more quickly while identifying the right place to go. Some sensors can also determine the sound of breaking glass, and differentiate between someone who has accidentally dropped a wine glass versus someone who has thrown a beer bottle in anger. In the latter case, the sound might be a leading indicator of a fight or worse. By dispatching police at the sound of the angrily shattered beer bottle, the subsequent melee or riot can be dispelled before it begins. In these ways, public lighting makes us safer yet again.

## POLLUTION

A city takes in resources and puts out waste. We call this the urban metabolism. It works just like a human body or an individual cell, which takes in food and generates waste.[21] That waste is a form of pollution. William Cronon, writing about Chicago, captured the idea this way:

> In any ecosystem, only the sun produces. All other beings consume in a long chain of killing and eating that stretches from the tiniest microorganism to the most aggressive carnivore. Since no organism can make energy, each must do its best to store it, accumulating a stockpile for use when the sun will not be so generous with its gifts. The same is true of human society: most of the labor that goes into "producing" grain, lumber, and meat involves consuming part of the natural world and setting aside some portion of the resulting wealth as "capital."[22]

These waste streams, or outputs from the city and the food, water, and energy systems that nourish it, are the pollution we need to deal with. And they are significant in scale.

In many depictions in popular culture, cities are dirty and
dangerous. Think of *Blade Runner* or *Soylent Green* or other dystopian movies that show cities as overcrowded and filthy. Their
urban metabolism is producing waste faster than the cities can
manage, and they have trouble bringing in the sustenance—the
food and water—that is needed to keep everyone healthy and
alive. The cities choke on their own waste. Despite these dystopian views, in reality, modern forms of energy enable cleaner
cities. The concentration of sewage in places like London and
Paris led to the development of sewer systems that later became
a fixture of any modern town. Similar tunnel systems move gas,
power, and water around our cities quietly and cleanly. In rural
areas, wastes could be managed more easily because their relative scale compared with the ambient environment is so much
smaller.

While many have fond, nostalgic impressions of the cleanliness of the good old days, urban pollution could be unbearable.
The ash and smoke from energy consumption was a defining
characteristic of London. Charles Dickens captured some of the
filth and pollution of London in his book *Bleak House*, which
was released serially in the 1850s but set in early 1800s London.
The book was ostensibly about the ways lawyers made money
from legal battles—a refrain that continues today—but along
the way presented an environmental criticism that notes the
extensive pollution of urban life at the time. Allen MacDuffie,
an expert on Victorian literature, noted that the novel can now
be understood as part of a nascent discussion of energy use and
pollution that emerged in the mid-nineteenth century, "during
the rise of a new resource-intensive industrial and economic
order."[23] According to him, Dickens's book is part of a wave of
Victorian novels that first articulated the environmental and
sustainability challenges of modern energy, including "the gap
between cultural fantasies of unbounded energy generation,

and the material limits imposed by nature." In other words, Victorian literature set the stage for the environmental movement that blossomed a century later to tackle some of the same challenges.

The opening paragraphs of *Bleak House* unfold as follows (emphasis added):

> LONDON. . . . *As much mud in the streets as if the waters had but newly retired from the face of the earth. . . . Smoke lowering down from chimney-pots, making a soft black drizzle, with flakes of soot in it as big as full-grown snow-flakes—gone into mourning, one might imagine, for the death of the sun. . . .*
>
> Fog everywhere. Fog up the river, where it flows among green aits and meadows; fog down the river, where it rolls defiled among the tiers of shipping and the *waterside pollutions of a great (and dirty) city.* . . . Fog in the eyes and throats of ancient Greenwich pensioners, wheezing by the firesides of their wards . . .
>
> Gas [lights] looming through the fog in divers places in the streets. . . . Most of the shops lighted two hours before their time.

The soot and ash from burning wood and coal darkening the skies, the muddy streets (no doubt mixed in with manure), and the water pollution give this story its initial framing. While the reference to fog is a cliché marker for London, that the fog settles in the eyes and throats and causes wheezing is a hint that it's actually smog—pollution from burning fossil fuels—rather than a purely meteorological event. Dickens also throws in a useful reference to the gas lights that dotted the city and had to come on earlier than sunset because of the smog blotting out the sun.

The streets *were* filthy. An image that accompanied an article in *The Illustrated London News* from 1865, bitterly complaining about municipal leaders' seeming inability to keep the roads clean, shows it.[24] The mud ruined clothes, damaged carriages, slowed foot- and horse-traffic, and was an all-around irritant. The article even bemoaned the waste of all the sewage and manure and other decomposing animal matter, which could be used as fertilizer in agriculture but instead was mixed up with dirt in the streets, exposing travelers and passersby to pathogens and noxious fumes. The article noted that the muck was a health problem, causing coughs and in some cases loss of life.

New York City streets were also famously dirty. Horse-drawn carriages were a common form of transportation. Today they are a tourist attraction for overpriced rides around Central Park, but 100-plus years ago, they were more widely dispersed throughout the city. Horsepower is a fun way to travel for a modern citizen who rarely sees horses, but when there are many horses, there is also a lot of manure, as the London article noted. The piles of manure caked on the streets were legendary, and the smell, the sights, and the risk of disease created a public nuisance.

Laying cobblestones or other masonry to prevent streets from getting muddy was an age-old tradition. It was also very expensive and labor-intensive. The production of asphalt and tar from crude oil in the late 1800s opened up the possibility of a quicker, cheaper way to pave streets. Unfortunately, runoff could create wastewater challenges—in some respects it was simply moving the pollution, not eliminating it—because it helped concentrate the location of the waste. However, it would remove the waste off to the side, away from immediate contact with pedestrians. At the same time that crude oil made paved streets a more practical reality, it also obviated the use of horses for transportation. A barrel of crude oil produces many components. Asphalt and tar are two important ones, but so is the kerosene discussed earlier, along with diesel and gasoline. Those last two fuels, when combined with the new internal combustion engines invented by Nikolaus Otto and Rudolf Diesel, created a new trend for automobiles in the early 1900s. The streets, paved with asphalt and tar from crude oil, became easier to keep clean at the same time that gasoline and diesel, also from crude oil, eliminated one of the biggest sources of mess: horse manure.

By moving to cars and vehicles powered by gasoline and diesel, the manure was moved off the streets, soot and ash were reduced, and cities became cleaner. At least, they did at first. Yesterday's solutions are today's problems, and while cars eradicated the problem of manure almost completely, as they grew in popularity, so did their tailpipe emissions. And their noise. Cars with mechanical engines, metal brakes, and loud horns create a cacophony of noise pollution to match the air pollution. As a result, cities are significantly louder than rural areas, which is one of the reasons why rural areas remain desirable as an escape from the city. The noise is not just a nuisance: researchers have determined that excessive noise has less-than-obvious health effects such as reduced staff performance, daytime sleepiness, increased occurrence of hypertension and cardiovascular disease, and impaired cognitive performance by students.[25]

The air pollution can be significant. After decades of transition to cleaner fuels after the publication of *Bleak House*, visible pollution—floating ash, for example—was practically eliminated. But invisible pollution—such as nitrogen and sulfur oxides that form acid rain—kept growing. In the 1950s, things got so bad from the tailpipes and smokestacks that there was an acid fog episode in London known as the Great Smog of London, a handy bookend to the Great Stink from water pollution a hundred years earlier. A combination of pollution along with a cool inversion created a blanket of smog that smothered the city for four days in early December 1952. Because the weather was cold,

Londoners burned more coal than usual to keep warm. Because the air was still, the pollutants were trapped over the city for multiple days. As coal was burned, the emissions accumulated to dangerous levels. The coal was a dirtier coal, which spewed more pollutants per pound than low-sulfur or cleaner coals that are more typically burned in the United States and UK today. Plus those famous steam-powered locomotives, which used coal to boil the water, and diesel-fueled red London buses added even more pollution.

Visibility was horrendous, limited to a few feet. The smog was also a killer. It was essentially a stagnant soup of sulfuric acid that hung over the city and seeped inside homes and businesses, penetrating the lungs of guiltless victims. Thousands of people died from respiratory problems—modern estimates suggest the total was 12,000 deaths from the four-day event.[26] At least another 100,000 people—and possibly 200,000—were sickened by it. Then, when the weather changed, the wind picked up and the smog dispersed.

California is another natural place for smog. Because of its love affair with cars and its geography in which a natural basin traps pollution, Southern California became the poster child for smog. Clever businesspeople sought to capitalize on this during the 1984 summer Olympics, which were hosted in Los Angeles, by selling empty bottles as a tourist's souvenir labeled "L.A. Smog." The smog was as symbolic of the region as the Hollywood sign. Because the state's air pollution was so severe, California led the way in tightening emissions standards for cars, leading to a multi-decade cleanup. The air is cleaner there today than it was decades ago, despite more people driving more cars more miles. This case study shows that it is possible to make environmental progress even with population growth and without a slowdown in activity.

## THE FIRST MODERN CITY

Chicago is a useful case study in the rapid growth of a city. Its population doubled approximately every decade from 1850 to 1890 and then grew more than 50 percent between 1890 and 1900.[27] It started with a population of less than 30,000 people in 1850 and had 1.7 million people in 1900, which is astonishing.

This population growth created new problems. As Martin Doyle wryly concluded, "getting water out of a city is as important to the growth of great cities as getting water in." That meant, before Chicago could grow that rapidly, it would have to fix its sanitation system. In the 1850s, Chicago used privy vaults (a form of urban outhouses), which were not suitable for the scale of wastes that would be generated by a thriving population alongside factories and slaughterhouses that created their own concentrated waste streams. Taller buildings concentrated people where they lived, and factories concentrated them even more so during the workday. Drinking water and sanitation systems had evolved but had not kept up with the demands that were placed on them. It was not until the New Deal era in the 1930s that updated solutions became widespread. The New Deal brought rural electrification and urban sewers forward almost simultaneously. Rural electrification was needed to improve the quality of life in the country, and sewers were needed to improve the quality of life in the cities.

Chicago became the first major modern city for a few reasons, one of which is that it had an opportunity to build anew. The massive fire of 1871 wiped out a significant swath of Chicago, giving it a clean slate at the same time new forms of energy enabled new schemes of construction and transportation. Trains in particular helped make Chicago what it is today, partly because of its location between the raw resources of the Great West and markets along the Eastern Seaboard. Historian William Cronon calls Chicago "Nature's Metropolis."[28] His thesis is that the stories of Chicago and the Great West are interconnected; you cannot tell the story of one without the other. The description for his book reads: "This is the story of city and country becoming ever more tightly bound in a system so powerful that it reshaped the American landscape and transformed American culture." Going further, he claims that "no city played a more important role in shaping the landscape and economy of the midcontinent during the second half of the nineteenth century than Chicago." The birth of a great city happened hand in hand with the availability of new energy sources and their companion technologies.

Chicago boomed partly because it was a transportation hub. Chicago is sited in the flats between two great waterways, the Mississippi River and the Great Lakes, thereby connecting trappers, farmers, and settlers in the nation's interior with the Atlantic economy.[29] Trains only enhanced those connections, putting Chicago at the center of a huge network of economic links that spanned the Midwest. Chicago became a natural hub for all sorts of commodities and finished products.

Trains, ships, and barges brought raw materials to Chicago, where they were upgraded into products and sent back out. Chicago depended just as much on the resources from the surrounding lands as those rural communities depended on Chicago for money and finished goods. It makes perfect sense that Sears and Montgomery Ward—the pioneers of mail-order catalogs—were both founded in Chicago. Chicago is also where agricultural commodity markets seemed to come alive.

The flow was nonstop. Timber in, lumber out. Timber (from the upper Midwest, including pines and coniferous trees, or from the eastern oak-hickory forests) flowed in via the lakes and rivers and was processed in Chicago sawmills and stored in some of the world's largest lumberyards before departing to prairie settlements in the West for fences and houses or to cities in the East. Pigs came in, and cut meats from some of the world's largest slaughterhouses went out via refrigerated railcars. Per Cronon's description, corn flows in like a continuous river of gold into the world's largest mills and silos before cornmeal flows back out. If it could have been viewed from above, it must have resembled a series of arteries and veins bringing materials in and out. At its peak, the Chicagoan empire, measured in terms of the expanse of its economic linkages, was as large as the Roman Empire.

But all of this intensification was impactful: remains of slaughtered animals (from the stockyards and slaughterhouses) and grease streaks on the ground, ash and soot in the sky "blanketing the prairie, this fall of filth, like a black snow." A quarter million trees would be cut down in a single year to satisfy the insatiable hunger for timber in Chicago. Cronon captures it this way:

> In nature's economy, all organisms, including human beings, consumed high-grade forms of the sun's energy—foods—and transformed them into low-grade ones. Although plants might convert the sun's energy into usable carbohydrates, and animals might then concentrate that stored energy in their flesh, they all finally drew their sustenance from the light of the nearest star. The abundance that fueled Chicago's hinterland economy thus consisted largely of stored sunshine: this was the wealth of nature, and no human labor could create the value it contained.[30]

He went further, observing: "Chicago's explosive growth was purchased at the expense of prairies and forest that had spent centuries accumulating the wealth that now made 'free land' so attractive. Much of the capital that made the city was nature's own."

The urban metabolism of a modern city was hungry for inputs, and those inputs came at a cost. Yet citizens were slow to stop the damage. Despite rampant deforestation, promoters and other prominent locals simply didn't believe it was possible to run out of trees. Companies pursuing timber had to move farther west and north to get trees because the easy-to-reach ones had already been cut down. Then they moved the trees by barge to Chicago. As they went farther away, the cost of wood increased, inviting an alternative. That alternative was coal. Because coal was closer to market (in mines in Pennsylvania or Illinois) and burned hotter, brighter, and cleaner, it was a higher-performing and affordable alternative. With the rise of coal, the deforestation slowed.

In 1882, the Chicago lumber market bottomed out about a decade before the World's Fair and the arrival of electricity, which would make wood less critical as a source of heat and light (though it would still be used as a building material). The World's Fair in 1893 was a turning point for Chicago. It helped announce Chicago's arrival onto the world stage while also giving it an opportunity to herald the arrival of modernity. The World's Fair established a vision for an urban life, skipping over the dirt, noise, and blood of the city's dirtier, grimier role converting natural resources into goods, to instead show a harmonious urban vision. The grounds included designs and layout by one of the most notable American architects, Frederick Law Olmsted, who designed Stanford University's campus and Central Park in New York City.

The buildings were gargantuan and all coated with white plaster and white lights. It was nicknamed the White City, and the flow of visitors—billed the White City Pilgrimage—had a daily attendance that reached 150,000 people. There were many exhibits and buildings, but one stuck out in particular: "The Manufactures Building would be the largest such structure on earth, covering twice the area of the Great Pyramid."[31]

Two other notable features made their debut at the White City. Desperate to find a breathtaking feature that would rival the Eiffel Tower from the Paris exhibition, the Chicago planners after much rumination settled on the Ferris wheel. This giant wheel was so christened for its inventor, George Washington Gale Ferris, but the name sounded like *ferrous*, almost like a hint of its metal construction and the age of steel skyscrapers. The spoke-and-hub design looked impossible, and critics did not believe it could work and sustain itself under its own weight. But it worked fabulously and was a spectacular global sensation.

The other prominent debut was electric lighting. The buildings and grounds were lit with 7,000 arc and 120,000 incandescent lamps, creating a spectacular display illuminating the night sky. It was also the culmination of a great rivalry between George Westinghouse and Thomas Edison, who had competed for the bid. Westinghouse's much cheaper bid based on alternating current (AC) was selected instead of Edison's direct current (DC) version. It helped demonstrate the viability of AC power, and in fact, all major grids in the world today are still built on that principle.

Chicago's emergence in the late 1800s exemplified the concept of urban-rural interconnectedness for mutual economic prosperity. The physical connections between the prairies and the skyscrapers were new. Today, modern cities are connected through global supply chains for food and other products and by information to other cities. Chicago used to be connected to the Great Plains by steam-powered trains and ships, but today it is more likely to be connected to London through a modern information supply chain powered by electricity. The physical supply chain of manufacturing and distribution is less important than the information supply chain. As one observer noted in follow-up work that builds on Cronon's writings, the information economy improves quality of life and wealth, but only for those directly participating. In the old model (urban area with hinterland) everyone participated in the economy, but the new model (urban area connected to other urban areas) leaves out many rural areas and people who don't have the skill set those modern industries desire.[32] This feeling of being left behind drives many political outcomes, such as the election of Donald Trump and Brexit. These modern political outcomes are a reminder that energy transitions are disruptive: they bring new advances but can leave many people behind.

## THE MODERN SMART CITY: REDUCED WASTE

Simply stated, cities are an intensification of population, which centralizes societies' energy usage into a reasonably localized system. That means fixing cities affords us the opportunity to use smaller-scale, local initiatives to take a huge chunk out of our energy expenditures at a scale that matters. Because cities consume such an incredible amount of energy inputs and generate a commensurate amount of wastes, they are where our energy conundrum of upsides and downsides is felt in full force. Using cities as a test bed to solve this conundrum is a prime opportunity. To do so, we should start by reducing waste. One approach is to make cities smarter so we can transform costly wastes into valuable resources, making cities highly efficient.[33]

On December 20, 2015, a mountain of urban refuse collapsed in Shenzhen, China, killing at least sixty-nine people and destroying dozens of buildings.[34] The disaster brought to life the towers of waste in the dystopian children's movie *Wall-E*, which depicted the horrible yet real idea that our wastes could pile up uncontrollably, squeezing us out of our habitat. One of the most powerful ways to transform an existing city into a sustainable city—one that preserves the earth rather than ruining it—is to reduce all the waste streams and to then use what remains as a resource. Waste from one process becomes raw material for another.

Cities are rife with wasted energy, wasted carbon dioxide, wasted food, wasted water, wasted space, and wasted time. Reducing each waste stream and managing it as a resource—rather than a cost—can solve multiple problems simultaneously, creating a more sustainable future for billions of people. Today, just flushing the wastes away isn't enough, however. After we reduce waste, we should close the loop and use the remainder again. First, limit waste, and then put it to work.

This new thinking begins when we redefine our concept of pollution. Raj Bhattarai, a longtime engineer at the municipal water utility in Austin, Texas, has proposed a new definition for pollution: *resources out of place*. Substances are harmful if they are in the wrong place: our bodies, the air, the water. But if they are in the right place, they are a resource. For example, instead of sending solid waste to a landfill and paying the bill, it can be incinerated to generate electricity. And the sewage from a million-person city can be mined for millions of dollars of gold and other precious metals annually for use in local manufacturing.[35] Thinking about waste in this way fits with the larger concept of the so-called circular economy—where society's different actions and processes feed into one another beneficially.

We can even use our waste $CO_2$ as an input for new materials. I often joke we need to invent a technology that takes $CO_2$ out of the atmosphere to create a novel, self-reinforcing material that gets thicker and stronger the more $CO_2$ it removes. It would be even better if this advanced material could be used as a feedstock for making products, as a building material, or as fuel. For the punch line I note that we could call this new device a "tree." In other words, wastes can be useful.

Simply put, waste is what you have when you run out of imagination.

One obvious place to start reducing waste is leaky water pipes. A staggering 10 to 40 percent of a city's water is typically lost to leaks in pipes. And because the city has cleaned that water and powered pumps to move it, the leaks are throwing away energy, too.

Energy consumption itself is incredibly wasteful. More than half the energy a city uses is released as waste heat from smokestacks, tailpipes, and the backs of heaters, air conditioners, and appliances. Making all that equipment more efficient reduces how much energy we need to generate, distribute, and clean up.

Refuse is another waste stream to consolidate. The United States generates more than 4 pounds of solid waste per person every day. Despite good efforts to compost, recycle, or incinerate some of it, a little more than half is still disposed of in landfills. Reducing packaging for consumer goods would help lessen this volume while also generating other benefits. Big retailers like Walmart, for example, have found that reducing packaging size not only eases pressure on the environment but also results in fewer trucks needed for shipping and more shelf space to display more goods.

Wasted food is its own heart-wrenching issue. Despite famine and food scarcity in many places globally, Americans throw away 25 to 50 percent of their edible food. Food is resource-intensive—requiring vast amounts of energy, land, and water to grow, produce, store, prepare, cook, and dispose of—so wasted food leaves a significant impact. Initiatives to reduce food waste that have popped up in the United States, such as the #IValue-Food campaign, and the United Kingdom are a start to resolving this problem.

Once cities reduce waste streams, they should use waste from one urban process as a resource for another. It is rare that waste is used so beneficially, but compelling projects are rising in certain places. Modern waste-to-energy systems like the one in Zurich, Switzerland, burn trash cleanly, and others like one in Palm Beach, Florida, recover more than 95 percent of the metals in the gritty ash that is left by the combustion.[36] Rural towns such as Jühnde in Germany create enough biogas from cattle manure to heat or power a large portion of in-city homes. My research group at the University of Texas at Austin has demonstrated that a cement plant in New Braunfels, Texas, can burn fuel pellets made of unrecyclable plastics rather than coal, avoiding $CO_2$ emissions and impacts from coal mining.[37]

Even trash that is landfilled can ultimately provide some value. Cities can collect methane that rises as the waste decomposes, which is an obvious improvement over flaring (burning off) the gas or simply letting the methane waft up into the atmosphere, where it traps much more heat than the equivalent amount of carbon dioxide. Power generators can convert the collected gas into electricity. Vancouver's landfills extract methane and burn it to heat nearby greenhouses growing tomatoes.[38]

Even then, landfills are still very leaky. The city of Austin makes methane from organic waste while producing 'Dillo Dirt that can be added to soil to increase its productivity, extracting even more value while avoiding more environmental impact. Thinking this way reduces the city's cost of managing wastes, too. These solutions solve multiple problems at once: saving money for energy that would otherwise have been purchased, reducing the need for expensive landfills, and avoiding unnecessary use and damage of land.

Wasted heat is another big opportunity. Harvesting waste heat is difficult because the low temperatures are hard to convert into electricity. NASA developed thermoelectric generators to do this on its spacecraft, but the technology is expensive or plagued with very low efficiency. Nonetheless, advanced materials that can more effectively convert heat to electricity are coming. One place to start the conversion is hot wastewater. When we wash our clothes, dishes, or bodies, the hot water goes down the drain. And the energy from that heat can be harvested. Sandvika, a suburb of Oslo, Norway, has massive heat exchangers along city wastepipes that extract the heat and use it to warm dozens of nearby buildings.[39] The city also uses the warmth to defrost sidewalks and roadways. Vancouver liked the idea so much it repeated the concept, using wastewater to heat thousands of condominiums and the Olympic village.

Taking that idea further is the Kalundborg Symbiosis in Denmark, a leading example of closed-loop thinking.[40] The industrial park has eight companies—centered around electric, water, wastewater, and solid waste facilities—that are interconnected such that the waste from one is an input for another. Pipes, wires, and ducts move steam, gas, electricity, water, and wastes back and forth to improve overall efficiency and reduce total wastes, including $CO_2$ emissions. Excess gas that had been flared off at the refinery is now sent to a drywall manufacturing plant in the park that makes the drywall with gypsum collected from the coal plant next door. Wastewater from the refinery flows to the power plant, where it is used to preheat the boiler feed or clean and stabilize fly ash from coal combustion. The refinery also sends waste steam from its processing to Novo Nordisk, which puts the heat to work growing about half the world's supply of insulin with bacteria and yeast. The pharmaceutical factory sends its agricultural sludge to nearby farmers. The coal plant sends its fly ash to a cement producer, waste heat to the biorefinery, and its warm water to an adjacent fish farm. And so on. The whole park looks like a living, industrial organism. And it has demonstrated economic growth with flatlined or reduced emissions.

Can the Kalundborg Symbiosis model be replicated on a larger scale, for cities worldwide? Yes, but only if we make cities smart. An industrial park is flexible because it has only a few tenants and decision-makers, but a city has millions of individuals making many independent decisions about energy, water, and waste each day. Integrating these disparate events requires either a major cultural shift—which can take decades to achieve, and without guaranteed outcomes—or advances in smart technologies to do the thinking for us. Or both. "Smart cities" are enabled by ubiquitous sensing and cheap computing power, compounded by machine learning and artificial intelligence. This combination can identify inefficiencies and optimize operations, reducing wastes and costs while enabling automatic operation of devices.

Thankfully, making cities smart is an alluring objective for planners seeking to accommodate higher densities of people without diminishing quality of life. For example, population and public health problems are severe in India, and in response Prime Minister Narendra Modi has announced an intent to convert 100 small and medium-sized municipalities into smart cities as a possible solution.[41] Clearly he sees cities as opportunities for improving the condition of Indian citizens.

The "smart" moniker itself is an accusation that most cities are dumb. The accusation sticks because cities rife with waste seem to be operating stupidly, with little information. They will need to get smarter. The US National Science Foundation has just launched a major research initiative called "Smart and Connected Communities," which is a sign that they have also identified cities as the leaders in solving these challenges. That phrase, by the way, indicates that intelligence isn't enough—interconnections among systems and people matter, too.

Smart cities rely heavily on big data, gathered from widespread sensor networks, and advanced algorithms to quickly gain insights, draw conclusions, and make decisions. Connected networks then communicate those analyses to equipment all across the city. Installing smart meters for closely tracking electricity, natural gas, and water use by time of day and household and industrial appliance is an obvious place to start. Real-time traffic sensors, air quality monitors, and leak detectors are also at hand. The Pecan Street consortium in Austin, Texas, is in the process of collecting data from thousands of homes to learn how access to more data streams like these might help consumers change their behaviors in ways that reduce consumption while saving costs. Cities like Phoenix, and military bases like Fort Carson, Colorado, have made pledges to become self-sufficient users of energy and water and net-zero producers of waste; achieving those ambitious goals will require a lot of interconnected data.

Making our infrastructure smarter is certainly the key to solving basic problems like leaky water pipes. Identifying leaks should be easy if meters are distributed throughout a water system; tracking flows becomes straightforward, so the amount and location of leaks can be pinpointed readily. Researchers in Birmingham, England, developed a system with minature pressure sensors that use a tiny amount of power to detect leaks in water networks nearly instantaneously, an improvement over the old technique of waiting for someone to call in and complain that water is shooting like a geyser out of the road.[42] And someday we might send smart robots down the pipes to repair the problems.

High-performance sensors will also let us find and predict natural gas leaks before accidents happen. Gas leaks are not only bad for the environment and a waste of resources, but dangerous, too, as we see in headline-grabbing explosions in urban areas with aging infrastructure. Modern cities with smart, self-healing infrastructure will detect leaks quickly or prevent them in the first place.

Turning profligate cities into places that reduce waste and reuse what's left will not be easy. Integrated R&D investments from the federal government will have to be combined with smarter policies from all levels of government and the technical innovation of private markets. Unfortunately, R&D funding is in recent decline, and in the United States it may drop further.

Investment has to be socially smart as well. Studies show that more R&D for smart cities has been focused on technology rather than what the citizenry needs.[43] Done the wrong way, the benefits of a smart city might accrue to those who already have Internet connectivity and access to advanced technologies, which would only widen the technology gap on top of other socioeconomic divides.[44] We need to help residents become smart citizens, too, because each individual makes resource decisions every time he or she buys a product or flips a switch. Access to education and data will be paramount. Connecting those citizens also requires urban landscapes that invite collaboration and neighborly interactions—parks, playgrounds, open spaces, shared spaces, schools, religious centers, and community centers—all of which were central tenets of centuries-old design for thriving cities. The more modern and smart our cities get, the more we might need these old-world elements to keep us together.

# Chapter 6

# SECURITY

Wars are often about resources. Sometimes one group facing scarcity fights with a group with abundance to get access to their resources. Sometimes resource constraints are a destabilizing force that makes it easier to recruit fighters. And, conflict itself can cause resource scarcity. Wars in the past have been fought over resources like food, water, or salt.[1] Over the past century, however, wars have been waged with and over energy, particularly oil. The story began in World War I, with its mechanized tank forces. The successful use of oil in that global conflict raised its strategic value, making it a resource worth fighting over.

For example, the Chaco War in the 1930s saw conflict between Bolivia and Paraguay for control over a region rich in oil. And the embargo by the United States and its allies on oil exports to Japan was the precipitating factor of the attack on Pearl Harbor in 1941. Iraq invaded Kuwait in the early 1990s to get its oil resources. The US response, along with an international coalition to help Kuwait repel Iraqi forces, was an intervention to keep oil flowing to global markets and to prevent Iraq from

controlling too much of the world's oil reserves.[2] President
George H.W. Bush specifically cited low oil prices as one of the
benefits of winning the war once Iraq was defeated.[3]

Looking forward, observers have wondered aloud whether
the next battleground for oil could be the Arctic.[4] Geologists
believe that there is a significant resource there, and several
countries—Canada, Denmark, Finland, Iceland, Norway, Russia, Sweden, and the United States—stake a claim to Arctic
territory. While exploring for oil so far north might put a chill
on international relations, climate change's great irony is that
by reducing the extent and duration of the Arctic sea ice, it has
made oil exploration easier.

Energy is a cause, weapon, and target of war. It gives us the
luxurious trappings and cozy comforts of life that make tempting targets for adversaries and gives us new weapon systems for
defending ourselves and attacking others. As energy has become one of the central issues of society, it has become one of
the central causes of global insecurity and key enablers of warfighting. Ultimately, modern forms of energy changed the way
war is waged. Done the right way, energy improves our security.
But done the wrong way, energy can worsen it.

## ENERGY AND THE WORLD WARS

The world wars fundamentally changed the role of energy in
warfare and defense. The success of energy in World War I was a
paradigm shift that accelerated through subsequent wars in the
twentieth century. World War I demonstrated the advantages
of oil for its improvements to land- and sea-based movement
of troops and weapons, while World War II demonstrated that
modern energy forms—namely oil and electricity—were indis-
pensable for advanced warfighting capabilities from the air.

According to Pulitzer Prize–winning oil historian Daniel Yergin, World War I was the first war to run on oil.[5] Armored tanks made their large-scale debut on the battlefield. Navies converted from coal to diesel, making oil, which until that point had been a consumer good, into a strategic commodity. Just as the internal combustion engine gave drivers the freedom to explore new places for work and tourism, it also freed up military forces. Prior wars fought with horses or trains were limited by feed or tracks. Trucks, tanks, and cars were more nimble than animals or the heavier, slower locomotives, giving Allied Forces—which had more vehicles—a military advantage and ultimately securing victory. Lord Curzon, Britain's foreign secretary, said, "The Allied cause had floated to victory upon a wave of oil."[6]

Much of the oil for these vehicles came from the United States, creating a transatlantic supply chain that was an easy target for German U-boats. Germans realized that targeting the supply chain of oil would disrupt the fighting capabilities of the Allied Forces in World War I, which is a logistical lesson US forces still struggle with in the Middle East.

Military strategists like to say, "Amateurs study tactics, professionals study logistics."[7] The adage holds true for most extended conflicts—and World War II is no exception. Strategic management of the production, movement, and consumption of energy—specifically petroleum, plutonium, and hydropower—was key to victory by the Allies.[8]

The US Navy's fleet expanded from just a few hundred ships in 1939 to 6,700 ships by the end of the war, of which 1,200 were auxiliary ships used for transporting fuel, ammunition, and food.[9] In the same six-year span, the United States produced nearly half a million aircraft, armored vehicles, and tanks, along with several hundred thousand bulldozers, half-tracks, trucks, and jeeps. Amid this massive industrial mechanized mobilization, the United States also managed to conduct the largest government development effort ever undertaken at that time: the Manhattan Project, which developed the atomic bomb in just three and a half years.

This explosion of production leading up to and during the war was not unique to the United States. Both Germany and Japan worked feverishly to supply their armies and navies with war material to counter the production of the United States and its allies, with a high degree of success. The Germans' technical and industrial prowess in designing and manufacturing was legendary, and was responsible for breakthroughs in armor, artillery, aviation, rocketry, submarines, and small arms. Likewise, the Japanese produced some of the most capable aircraft and ships of the era. Even though Germany and Japan made great strides in arming and supplying their forces, the United States had two key strategic and logistical advantages: oil and water.

Because World War II was so heavily mechanized and, unlike World War I, fought on a truly global scale, powering and supplying such immense and globally disparate military capability required vast amounts of energy—something the United States, as the world's largest petroleum producer at the time, had in much greater abundance than Germany or Japan.

The outcome of World War II depended even more heavily on oil than that of World War I. In addition to the widespread use of mechanized fighting machines such as tanks, World War II was also the first actively airborne war. That meant energy— for making and fueling the planes—played a bigger role than in any prior conflict.

Powering the broad range of ships, planes, and land vehicles with their equally broad range of petroleum-based fuels and lubricants required millions of barrels of oil. Transporting people, munitions, food, and fuel around the planet to multiple theaters of operation required even more. Fortunately, in the early twentieth century, the United States had begun an oil boom. And for enough oil to fuel the war, the country looked to East Texas.

The Spindletop oil discovery in far eastern Texas during the early half of 1901 marked the beginning of the Texas oil boom. The 1930 Bradford discovery in Rusk County, also in eastern Texas, became known as the significant East Texas Field. By the summer of 1931, more than 1,200 wells were in production, supplying 900,000 barrels per day.[10] Between 1935 and 1945, in order to support war efforts, annual production increased significantly.

While today we worry that the oil supply from the Middle East to the United States will be disrupted, in World War II the only threat to the security of domestic US oil was associated with the transportation of Texas oil to the East Coast states via the Gulf of Mexico. In 1942 and early 1943, several oil tankers were sunk by German U-boats off the East Coast. To circumvent this risk—in a feat that would be hard to duplicate today because of opposition from environmentalists or land rights groups—the United States built a nearly 1,500-mile-long pipeline known as the Big Inch between the East Texas fields and the nation's major refineries and distribution hubs in Philadelphia. The project was completed in less than twenty months, between August 1942 and March 1944. The pipeline carried more than 350 million barrels of oil by the end of the war. It is hard to imagine building such an extensive pipeline today so quickly. For example, the Keystone XL pipeline that could move oil from the sands of Alberta, Canada, to the refineries in Texas has been caught up in regulatory or legal battles for a decade.

In contrast to this secure, cheap, and nearly limitless supply of oil in the United States, Germany and Japan had to fight for and import nearly all of their oil. The Germans were heavily dependent on the Soviet Union and Romania, drawing 75 percent

of their oil supplies from them, which was one of the main reasons for Nazi Germany's nonaggression treaty with the Soviets
at the start of the war. It also explains some of Germany's east-
ward movements during the war, as they sought to gain access to
oilfields to fuel their military machine. And in response to the
difficulty in securing the supplies of petroleum they needed in
the face of embargoes and relentless attacks by the Allies on oil
supply lines, Nazi Germany started an ambitious program to turn
abundant domestic coal into liquid fuels. The coal-to-liquids
program was a technical success, but it is a little bit like squeezing blood from a stone (in this case, oil from a rock): it was
costly and diverted manpower, undermining German warfighting effectiveness. The only other country since then to attempt
a large-scale effort to convert coal to liquid fuels was South Africa toward the end of the apartheid era, when a global embargo
cut off their access to oil from the world markets, after which
they turned to their abundant domestic coal resources. When
push comes to shove, a country will try to use coal in place of
petroleum, but petroleum is much cheaper and easier.

Likewise, Japan—which has no domestic coal reserves and
was not aware of the German technology for converting coal
into liquid fuels anyway—had relied on the United States for
about two-thirds of its oil in 1939, while British territories and
Dutch India supplied the rest. According to one analysis, "Japan's position was so vulnerable that its lifelines were in the
hands of the Allied countries, particularly the United States."[11]
The United States had been shipping oil and aviation fuel to
Japan from the West Coast under a 1911 trade agreement. After
Japan invaded Indochina in 1941, however, the United States
issued an oil embargo against Japan, which was joined by American allies. Months later, Japan retaliated with the attack on
Pearl Harbor. Throughout the war that followed, Japan relied
almost solely on oil from the Dutch East Indies. This primary

source was almost completely shut off by the end of the war through the success of American submarine attacks on Japan's oil tankers. The American victory over the Japanese is often said to hinge on the atomic bomb, but it also had a lot to do with the way the United States undermined energy supplies to Japan.

During World War II, the United States built an armada of more than 300,000 advanced aircraft. That would not have been possible without enormous quantities of aluminum. And that, in turn, required enormous amounts of electricity and thus water.

In 1939, the United States produced nearly 164,000 tons of aluminum. By 1943, that production had quintupled to more than 920,000 tons.[12] Obtaining the bauxite ore to support this expansion in production was not a problem because there were ample sources throughout the country. Producing the electricity necessary to turn it into aluminum was a different matter. The difference between steel and aluminum is that steel is smelted with heat (usually from coal) and aluminum is smelted using an electrolytic process to separate aluminum from alumina, requiring 10 kilowatt-hours of electricity for every pound produced. Consequently, in the early decades of the twentieth century, America's principal aluminum producer (Alcoa, or the Aluminum Company of America) had been building not only smelting factories but also hydroelectric dams to provide electricity to them. At the time, electricity was also produced by burning coal, but hydroelectric dams could be much larger and generate more power in one spot, could do so cheaply, and were usually far from city centers so large consumers like aluminum plants had less competition for the electricity.

By the mid-1930s, Alcoa had built three dams on the Little Tennessee River as well as two in western North Carolina. The three dams in Tennessee alone provided a combined 265 megawatts of power to smelters in Alcoa, Tennessee. With the outbreak of World War II and the increase in demand for aluminum, Alcoa joined forces with the Tennessee Valley Authority to obtain even more hydroelectric power to run its rapidly expanding smelting facilities. But even this build-out was not producing enough aluminum to meet the insatiable demands of the military.

So Alcoa looked west to the hydroelectric dams on the Columbia River in Oregon and Washington. It built a string of factories between Longview and Spokane, Washington, powered by dams such as the Bonneville and Grand Coulee. The hydroelectric power plant at the Grand Coulee dam in particular, which generates 6.8 gigawatts of power today, equivalent in size to about six nuclear power plants, is still the nation's largest power plant.

The proximity of these Columbia River–powered aluminum smelters to the growing aviation industry of the Pacific Coast was helpful. The bulk of the aluminum from these plants traveled by rail along the West Coast, either north to Boeing's aircraft plants on Puget Sound or south to the plants of Douglas, Consolidated, North American, Curtis, Lockheed, Northrop, Hughes, and others in the Los Angeles Basin (all powered with electricity from the Boulder [now Hoover] Dam on the nearby Colorado River). Wartime aluminum production peaked in 1943 at 920,000 metric tons in a year; aircraft production peaked in 1944 with 96,318 planes. In 1943, the aluminum industry became the single largest consumer of electricity in the country, requiring 22 billion kilowatt-hours annually, a little over 8 percent of national energy consumption. [13]

This dramatic growth in the aluminum industry happened in tandem with another engineering feat: the development of the atomic bomb. By 1944, the nation's second-largest consumer of hydropower was an industry that hadn't even existed in 1943 and that also happened to be the nation's most highly guarded secret.

The Manhattan Project formed under a cloud of uncertainty. No one knew how to build an atomic bomb or even if it could actually be done. As a consequence of this uncertainty and the exigency of the effort, under fear that the Germans might succeed before the United States, two different design approaches were pursued. One used the uranium-235 isotope ($^{235}$U) as the fissile material, and another used plutonium-239 ($^{239}$Pu).

Because $^{235}$U constitutes only about 0.7 percent of naturally occurring uranium, with the other 99 percent being unusable $^{238}$U, huge amounts of uranium would have to undergo an enrichment process to separate out the less common, desired portions of $^{235}$U. Similarly, plutonium, which is more useful for atomic weapons, doesn't exist in nature and would have to be produced artificially in nuclear piles (the predecessors to nuclear power reactors) that had yet to be designed or built. Both options would require electricity and water.

Instead of choosing just one approach to obtaining $^{235}$U, the Manhattan Project leaders decided to simultaneously pursue three different methods of uranium enrichment. The three most promising methods—gaseous diffusion, electromagnetic separation, and thermal diffusion—would all require a great deal of electricity. Just like Alcoa, the Manhattan Project turned to the power of water and began construction of what would become Oak Ridge National Laboratory just west of Alcoa's headquarters in Tennessee. By the end of the war in 1945, Oak Ridge, which did not even exist in 1941, had grown to a population of 75,000 people and was Tennessee's fifth-largest city.[14] The main enrichment facility at Oak Ridge was built in 1944. It was more than a mile long, end to end; covered 44 acres; and had 2 million square feet and four floors filling an enclosed volume of more than 97.5 million cubic feet. It was the world's largest building.[15] Two more facilities for separating uranium isotopes were also built, and at their peak of enrichment activities, the three enrichment facilities were consuming 1 percent of all power produced in the nation.[16]

But even this level of effort could not produce enough highly enriched $^{235}U$ for more than one uranium bomb. The second path to a nuclear weapon, plutonium, proved to be a good bet. To produce large amounts of plutonium, the Manhattan Project engineers planned to build three nuclear reactors specifically designed to enhance the production of $^{239}Pu$ as a by-product of fission between $^{235}U$ and $^{238}U$. Because the plutonium would have to be chemically separated, they also needed two large facilities to handle the material, and because it is very radioactive and can accumulate in bones, working with it is dangerous.

Again, the Manhattan Project followed Alcoa's lead, and this time they turned to the Pacific Northwest. A remote site of more than 2,000 square kilometers was chosen in southern Washington due to its proximity to the Columbia River for cooling water and the Bonneville and Grand Coulee Dams for electrical power. Known as the Hanford site, it is today also the location of the US Department of Energy's (DOE) Pacific Northwest National Laboratory, which continues to produce significant scientific and technologic breakthroughs. That today's national laboratories are at sites next to dams because of the role of water in enrichment is a lasting testament to the strategic importance of electricity and water for generating it. They also affirm the notion that tomorrow's breakthroughs require a lot of reliable energy to figure out.

The three reactors were placed on the banks of the Columbia River for access to cooling water, at several kilometers apart for safety reasons. The reactors were designed to produce plutonium, not power, and consequently, the enormous amount of heat they generated in the fission process had to be removed. During operation, each of the three reactors required a continuous flow of 75,000 gallons per minute of the Columbia's chilly water.[17] The two enormous separation facilities also needed cooling and electricity, both supplied by the Columbia River.

The atomic efforts ongoing at Oak Ridge and Hanford came together at Los Alamos in New Mexico (now the location of DOE's Los Alamos National Laboratory, another world-leading center for innovation). The production of the Little Boy uranium bomb used at Hiroshima and the Fat Man plutonium bomb dropped on Nagasaki would not have been possible in such a short time period were it not for the abundant water and hydroelectric resources of the Tennessee Valley and Columbia River.

Meanwhile, Japan did not have a nuclear program (and has limited hydroelectric resources, anyway). Germany pursued a nuclear program, but because of the limited hydroelectric resources within its borders, the country had to look northward to Norway for abundant, cheap electricity (and for heavy water, or deuterium, to serve as a moderator for nuclear chain reactions), complicating its efforts to conduct a major weapons program in a secret and invulnerable way.

In the end, success in World War II was due to more than just abundant and reliable sources of oil and water in America. However, if it had not been for these natural resources, the speed and scale of the mechanization of the US military and its allies would have been significantly reduced, particularly in the early years of the conflict. The fact that these resources did exist and were already being tapped and utilized in the 1930s allowed for their rapid expansion prior to US entry into the war and contributed significantly to the conduct of the war—and its ultimate outcome.

The experience of World War II's Manhattan Project offers lessons about how to detect other countries' nascent nuclear programs. Because the electricity requirements for uranium enrichment are so intensive, tracking power demands is a useful way to monitor the progress of nuclear programs in countries like Iran. Nuclear power plants need the $^{235}U$ isotope to be enriched from its natural abundance of 0.7 percent to a higher concentration of approximately 5–20 percent for reactor-grade applications. However, weapons-grade fuels use highly enriched uranium with the $^{235}U$ isotope concentrated up to 85 percent of the mass. Energy consumption is a revealing indicator of the extent of the nuclear fuels enrichment process and therefore is of keen interest to international weapons inspectors.

## ENERGY AS A TOOL OF WAR

Energy does so much during war. For World Wars I and II, energy's primary value was to make and fuel the aircraft and land-based machines. Energy powers the communications systems, collects information, makes the weapons, and guides the weapons. It has reshaped how we make war and how we think about war.

During the Cold War, energy as a propellant and constituent of nuclear-tipped missiles gave modern arsenals a global reach with horrendous destructive potential. Those missiles' use (or lack of use) was governed by a policy of mutually assured destruction, the idea that the world's two great rival superpowers—the United States and the Union of Soviet Socialist Republics—would restrain themselves from launching a nuclear missile because doing so would unleash a counterattack on such a grand scale that it would ensure the destruction of each country. That energy could at once improve a nation's security and raise the risks of catastrophic attack are symbolic of energy's double-edged benefits.

There is also a whole class of systems known as "directed energy weapons" (DEWs). These weapons use a laser beam to inflict damage at a distance. LASER is an acronym for *light amplification by stimulated emission of radiation*, which is a tightly focused beam of light that has one color and points in one direction. These features contrast with a typical light bulb, which puts out light with many colors (white light is actually a combination of all the visible colors) pointed in all directions. By harnessing energy in such a specific, precise way, military personnel can harm other fighters by burning them, disable weapons such as incoming missiles, or blind the optical equipment of vehicles. DEWs also do not make noise when activated. They can also be designed to use invisible wavelengths and are not affected by wind, the Coriolis effect, or gravity the way traditional artillery is. There is even a version of a DEW called an electrolaser that ionizes a pathway through the air between the initiation and the target, which can then be electrified to form a bolt of intentional lightning, like a Taser, only with a range of several miles. These types of weapons are appealing because they can be used in a nonlethal way, but they are controversial because of their cruel ability to blind people.

Ironically, energy also enables greater field vision. One aspect of modern warfare is "eye-in-the-sky" surveillance capabilities from airplanes, satellites, and unmanned drones. Early planes were propeller-driven, which limited their altitude. Flying behind enemy lines exposed the pilots to the risk of getting shot down. Modern energy forms brought forth jet fuel and better materials, both of which enabled high-altitude airplanes. Planes made of lightweight aluminum were stronger and could fly higher. The metals also allowed for airtight plane construction, which meant the cabins could be pressurized and pilots could survive even at heights where the air was thin.

These features were deployed for spying as the US defense establishment pushed for more advanced planes that could fly even higher with powerful cameras, out of view from ground-based observers and—hopefully—out of reach from ground-based antiaircraft guns. The most notable example of this was the Central Intelligence Agency's (CIA's) famous U2 spy plane, which was developed to look for missile silos or other military installations in the Soviet Union. It was capable of flying at a height of 70,000 feet, about twice as high as commercial airlines. And, interestingly enough, coal plays a role in its fateful story. In an event that had global consequences, on May 1, 1960, American U2 pilot Gary Powers was shot down while spying on the Soviet Union. The Soviets' ability to shoot down aircraft at that altitude caught the Americans by surprise—unbeknownst to the United States, the range of their weapons had been extended significantly. This crash created an internationally embarrassing incident for President Eisenhower because it revealed he had lied about US spying operations. It also made Powers a household name. The reason Powers was there in the first place was because his father, Oliver Powers, had been a coal miner in Appalachia. He had vowed to himself that his son would never work in a mine. So he became a pilot instead.[18]

High-altitude planes clearly were not high enough. So we pushed even higher, ultimately launching satellites that could orbit the earth hundreds or thousands of miles above the surface while collecting information without the risk of getting shot. I learned about spy satellite programs from Dr. Hans Mark, who in the late 1970s ran the National Reconnaissance Office (NRO), which is one of the main US intelligence agencies and oversees many of the nation's spy satellites. He subsequently served as deputy administrator of NASA, where he was in charge of the space shuttle program, before his selection as chancellor of the University of Texas system. In that role, he served as my mentor and thesis advisor while I was an undergraduate student.

Without rocket fuel to reach escape velocity from the gravitational pull of the earth, these satellites would never reach their orbital destination. Without onboard power—either with long-lived nuclear systems, fuel cells, or solar panels coupled with batteries—the satellites would not be able to perform their tasks. Energy is critical to every part of their missions. Satellites are conveniently out of reach from ground-based antiaircraft weapons, but that means they are very far away. As a result, despite high-powered optics, they have limited resolution for the images they collect. And they need to either stick to their orbits, which limits their areas of surveillance, or spend precious fuel maneuvering to another orbit.

That is where drones can fill the void. Drones combine advances in communications, enhanced computational capabilities, and new, lighter materials, all powered or made by energy. Drones are appealing because they do not include a pilot. As such, they can fly longer (most pilot-based missions are limited by how long a pilot can fly before needing a bathroom break or a nap). And because they do not need to provide life-support systems such as pressurized air or a cockpit, they can be much smaller and lighter, saving energy, which increases endurance. That also means at high altitude they are hard to detect. Using high-powered optics derived from satellite-based systems but from a height of 60,000 feet, which is what the Global Hawk surveillance drone can reach, gives much more detailed information than can be gleaned from satellite images taken in low earth orbit.

The weaponization of drones was inevitable. The first prominent attacking drone was the Predator. I remember watching a real-time surveillance and attacking action from Nellis Air Force

Base, Nevada, in 2005. From this base, drones from halfway around the world are operated remotely by pilots out of harm's way. I felt like I was watching a video game as members from international coalition forces used joysticks to control the drone's flight, its optics, and its weapon systems, with displays on large screens covering a wall like something out of a movie. Our daytime visit to Nellis AFB coincided with late-night covert action in Afghanistan. I could see in real time as the drone scanned the countryside, following a car with insurgents who had no idea they were being followed by a drone tens of thousands of feet in the sky, being watched by pilots and analysts thousands of miles away. This was the ultimate in standoff distance.

Standoff distance is what lets you attack adversaries from afar. The farther the range of your capabilities, the easier it is to attack them while being out of range of their weapons. This is why the English longbow, catapults, rifles, and ultimately missiles are so useful: they all have greater standoff distance than pistols, muskets, or swords. Energy is the ultimate enabler of standoff distance because it gives propulsion over longer distances (as for planes and missiles) and allows remote sensing and operation (by means of satellites and drones). While this distance improves troop safety, it removes operators/shooters from the battlefield, which might have moral implications for the rules of engagement.

Energy also is key to targeting systems. As noted earlier, lasers are their own form of directed energy weapons, but they can also direct other weapons. The weapons carried by airplanes and drones have evolved over time from simple gravity-directed bombs, which fall when they are dropped, based on the wind speed and the plane's speed, to smart bombs, which have guidance capabilities such that they navigate to a specific location. Ultimately, laser-guided weapons were also developed. For these weapons, a plane or drone flying overhead has a targeting pod with optics to collect information. A laser pointer puts a dot on a target, and a guided bomb then maneuvers itself to the laser spot. That laser pointer can be carried by ground forces and has a range of a few miles. So someone on the ground, who might have a better view of where the bad guys are than a pilot thousands of feet in the sky, uses his or her laser pointer to put a spot on the building, which guides the bomb dropped from the plane flying overhead.

## ENERGY USAGE AS A WEAPON

Energy has become so fundamental to the proper functioning of societies that its economic importance has also become weaponized. It can be used to inhibit a rival's capabilities or to extract policy concessions. Most famous is the oil weapon, which is the intentional deployment of oil supply cutoffs. The oil weapon is used by producing nations to inflict economic pain on consuming nations, embarrass an importing nation, or achieve desired policy goals. In the conventional scheme, limiting the supply of oil to importing nations creates scarcity, which drives prices up, causing popular discontent.

An archetypal example of the oil weapon was the 1973–1974 oil crisis known as the Organization of Petroleum Exporting Countries (OPEC) oil embargo. Arab oil-producing nations angry at importing countries who supported Israel reduced their oil output by about 5 million barrels per day (about a 25 percent drop in production from those countries) and fixed prices 500 percent above their prior levels.[19] Ultimately, the international oil markets suffered a shortfall of 4.3 million barrels per day, and those oil constraints led to gasoline scarcity and high prices in places like the United States. The subsequent oil disruption in the late 1970s from the Iranian Revolution was also impactful. The two crises caused long lines at gasoline stations and high energy prices that pinched pocketbooks, inconvenienced drivers, and became a defining cliché of the 1970s. The crucible of those experiences forged a lasting memory that defined oil security concerns for decades afterward.

The cutoffs had a global impact with a cascading series of reactions. In the United States, the 1970s oil crises spawned the creation of the Department of Energy and new agencies for tracking energy statistics. The United States also increased federal research spending for alternative energy options, established corporate average fuel economy (CAFE) standards, and launched public information campaigns focused on conservation and efficiency, all of which dovetailed nicely with the burgeoning environmental movement that emerged in the late 1960s and early 1970s. Taking a page out of Germany's energy security playbook from World War II, a federal program was created to explore the possibility of converting domestic coal resources into liquid fuels. Ultimately, the project's high costs caused public support to wane, and it was killed.

In Europe, countries also implemented fuel economy standards and aggressively ramped up exploitation of different forms of energy. Starting in 1974, France in particular turned toward nuclear power as an alternative to oil and gas in the power sector. Consequently, France remains a world leader in technologies for nuclear power plants and reprocessing nuclear waste. About 75 percent of French electricity is generated from nuclear sources.

Oil supply cutoffs could remove anywhere from 1 to 6 million barrels per day from the global market of about 90 million barrels per day. That doesn't sound like much, but small changes in availability have outsized impacts on price. In response to the 1973 oil embargo and out of fear of future supply cutoffs, the International Energy Agency (IEA) was formed in 1974 to help coordinate international actions on shoring up supply.[20] In total, sixteen countries were original members of the IEA, and its primary mandate was related to international cooperation around energy security. Emergency response to oil supply disruption is still one of the main pillars of the IEA. Membership requires that countries hold oil stocks equivalent to at least 90 days of net oil imports and that they help other IEA members in the event of an oil crisis.

As part of its mandate, the United States developed the Strategic Petroleum Reserve (SPR), which stores more than 700 million barrels of oil in belowground salt caverns. These hollowed-out caverns in places like Texas and Louisiana are thousands of feet tall: two Empire State Buildings could be stacked one on top of the other inside the cavern underneath the ground's surface. The intent of the SPR is to stockpile oil to blunt the shock of future disruptions and to make sure a minimum amount would be available for emergency or military actions in the event of a major supply cutoff. Other countries did something similar. The risk hasn't gone away, so the IEA still publishes reports and actively engages countries from around the world.

It's not just oil that can be used as a weapon. As the IEA noted, "Energy security is no longer just about oil."[21] Natural gas and coal can also be cut off strategically. Decades after the 1970s energy crises, the oil trade has become a global market, which makes it easier to find alternative suppliers if one cuts you off. But the global natural gas market is nascent. That means some of the risks of oil security have shifted to natural gas. In some ways, natural gas might be an easier weapon to control, as its flow can be interrupted just by turning off a pipe rather than running a large-scale blockade. Oil is easy to move by pipe, truck, or ship. If one ship fails to deliver, another ship can be procured, and if a pipeline ruptures, oil can be delivered by truck. By contrast, gas is less fungible, so it's not possible to rapidly replace gas supplies: the pipes take a long time to build; because they operate at higher pressure, the natural gas pipelines are harder to repair than oil pipelines; trucks can't move very much gas; and liquefied natural gas (LNG) shipping requires specialized liquefaction or gasification facilities that take billions of dollars and several years to build. Moreover, the LNG ships are specialized and expensive. As a result, landlocked nations or those that have not already invested in LNG import facilities depend on natural gas shipments by pipe from production zones. That means the gas infrastructure in its current configuration is less resilient overall, which puts the producers in a position to control the flow of gas.

Because of pipelines that physically connect a supplier with a buyer, both have fewer options in the event of a dispute. And major suppliers like Russia have used that leverage in their favor. On multiple occasions since 2000, Russia has cut off natural gas supplies that travel through pipelines in Ukraine toward consumers in other European nations. The risk of losing access to Middle Eastern oil in the 1970s gave way to the risk of losing access to Russian gas in the 2000s and 2010s. And these disputes, cutoffs, or threats of cutoffs often occurred during the winter, when Europeans used gas for heating. One dispute in 2009 lasted three weeks during a bout of cold weather, leaving European consumer nations—especially Germany, a major importer of gas—at risk of being stranded in the cold.

Even fuels like coal, which are considered more secure because of their domestic abundance, can present risks. Following multiple disputes over several years and amid escalating tensions over Russia's invasion of Crimea, in November 2015 Russia cut off natural gas and coal supplies to the Ukraine.[22] Ultimately, American coal was exported to Ukraine to reduce its dependence on Russia and to make up the shortfall.[23]

The security relationship is not just about using oil as a weapon—it's also about the use of oil and weapons as negotiating tools. Energy producers use their oil as leverage to get weapons from the United States, and importing nations use weapons as leverage to get oil. In 2008, global oil prices were very high. The Bush administration was coming to an end, a presidential election was afoot, and the global recession was just starting. After a meeting with King Abdullah of Saudi Arabia, during which President George W. Bush asked that they increase oil production to reduce prices for American consumers, the headlines blared "Saudis Rebuff Bush's Request for More Oil Production."[24] In response, Democrats in the US Senate introduced legislation to block the sale of more than $1 billion of weapons to Saudi Arabia unless they increased their production by 1 million barrels per day.[25] Just as the Saudis had threatened oil cutoffs as a weapon, US policymakers used the threat of weapons cutoffs as a way to get oil. Ultimately, the flare-up settled down, and the Saudis stuck to their plan. In the battle between oil and weapon systems, oil has the upper hand.

Bush's entreaties came at a sensitive time. Energy security was not a national priority until the 1970s oil crises made it so. For the next few decades, the view on oil security focused on security of supply. But after the terrorist attacks of September 11, 2001, in the United States, the view on energy security evolved. Instead of worrying about steady supply, Americans worried that their oil purchases were funding terrorism. In Iraq and Afghanistan, we were fighting against groups funded by revenues from oil the West had bought. It was as if we were funding both sides of the war. Since fifteen of the nineteen hijackers on 9/11 were Saudi, that relationship was especially fraught. Asking the Saudis to increase oil production less than a decade after those attacks and while the US military was deeply engaged in the

Middle East highlighted the murky interconnections between energy and national security. And almost exactly a decade later, in an eerily similar replay of Bush's request, and despite the shale revolution and other major shifts in the energy sector, President Donald Trump also issued an entreaty to Saudi Arabia to pump more oil as a way to reduce global oil prices.[26]

Despite the ongoing attention to oil security, there's also a reason to believe the oil weapon would not work as effectively today as it did in the 1970s. At that time, and earlier, countries arranged bilateral export-import agreements. That is, an oil-importing country would arrange to buy oil from a particular oil-exporting country. So the selling country could hike the prices and cut off supply, leaving the importing country with few options. But since the 1980s, global oil markets have been liberalized, which means each purchaser has many options for suppliers and the price can fluctuate. In today's market-driven mindset, economists argue that the tactic of cutting off supply might not gain traction, because oil is now a fungible commodity. Buyers could simply obtain oil from another country or company if a supplier cuts them off. A producing country cannot prevent a consuming country from buying oil on the world market without military intervention such as a blockade.

Another reason why the oil weapon might not work anymore is that oil-producing nations need the money. The headline of an article in the *Wall Street Journal* in early 2007 asked, "Can Venezuela, Iran Endure an Oil Pinch?," rightly noting that producing countries might need the money from oil sales more urgently than consuming countries need the oil. That rationale is one of the motivations behind sanctions as a foreign policy tool with Iran and other countries. When their oil sales are cut off, the countries lose important revenue, inhibiting their ability to develop dangerous weapons systems. The jury is still out on whether that approach is effective in the long run, though.

Despite that rationale and decades of market liberalization, the oil weapon lives on today in one form or another. And it's not just "hostile" countries like Russia or those in the Middle East that consider oil an important negotiating tactic: Canada has also threatened to use the oil weapon. An article covering the 2008 US presidential election captured this incident as follows: "Canadian Trade Minister David Emerson mused that Ottawa might wield oil exports as a bargaining chip if NAFTA is reopened—as US Democratic presidential hopefuls Barack Obama and Hillary Clinton apparently want. Veiled threats about Canada's energy riches have been used sporadically over the years in the face of US protectionist threats."[27]

Two incidents in the span of one week in August 2009 involving Muammar Gaddafi, the longtime leader of Libya, illustrate the oil weapon's relevance long past the oil crises of the 1970s. In short, by cutting off oil to Switzerland and negotiating an oil exploration and production deal with the UK, Libya showed it was unafraid to use oil as a bargaining chip, and consuming nations showed they could be manipulated by it. One article that month noted, "Last week was a good one for Libyan leader Muammar Gaddafi. He not only succeeded in bringing home to a hero's welcome a terrorist who killed 270 innocent people over Lockerbie, Scotland, he also brought the world's least belligerent nation to its knees." It concluded, "36 years after the Arab oil embargo, the oil weapon is still alive."[28] In a controversial move that week, Gordon Brown's government in the United Kingdom released the Lockerbie bomber, Abdelbaset Ali Mohmed al-Megrahi, so he could return to Libya, in exchange for a multi-million-pound oil exploration deal with BP (formerly known as British Petroleum). One news article covering the story skipped the niceties, declaring matter-of-factly, "Lockerbie Bomber 'Set Free for Oil.'"[29]

In parallel, Libya used its oil to manipulate policy in Switzerland. After Gaddafi's son Hannibal was caught beating his employees, the Swiss jailed him. Gaddafi demanded his release and a public apology. While complying with the first demand, the Swiss did not apologize. Libya supplied 20 percent of Swiss oil consumption at the time, which gave the country leverage. Gaddafi cut off oil supplies, withdrew money from banks, and cut off diplomatic ties. The Swiss, eventually buckling under the pressure of many months of supply disruption, apologized in August 2009.[30]

But oil-consuming nations are not completely without recourse. Consumers can use conservation to cut off demand, dropping oil prices and inflicting economic pain on producers. It has been argued that strong US production after 2008, together with flattened consumption (because of higher fuel economy standards and shifts toward mass transit), significantly reduced US imports of crude from the world markets, helping to bring Iran to the negotiating table over nuclear energy.[31] Sustained low oil prices have also been partially credited for the collapse of the Soviet Union in the early 1990s. For countries like Russia and Saudi Arabia, oil revenues are a key influx to the federal treasury, so low oil prices or low production bankrupts those countries. In that way, the oil weapon's opposite—the oil conservation weapon—can be effective at taming some producing countries' worst instincts. Along the way, oil consumers save money on energy costs and live with less pollution because overall consumption is lower.

## SECURITY OF ENERGY SUPPLY

Up until the 1970s, oil and the mobility it enabled were considered an aid to national security. Oil enabled ships and planes with longer ranges and swifter action, both of which were beneficial. But the two energy crises in the 1970s changed that. Oil—imports in particular—became a national security liability. These crises created an enduring concern about the security of supply. After the attacks of September 11, 2011, energy security centered on the notion that energy purchases should not support countries that hate the United States. The response in the 1970s was primarily about fuel switching and conservation. The response after 9/11 was primarily to exploit domestic sources.

The energy security problems of the 1970s actually began decades earlier. The United States was a dominant oil producer for several decades after Spindletop, the spectacular gusher drilled in East Texas in 1901. US oil production fueled successful efforts in World Wars I and II. But after World War II, US consumption grew faster than production. Consequently, we turned to other regions of the world to import oil, and thereafter maintaining a steady global supply for Western allies became a national policy priority.

The United States has also made clear that access to oil was something worth going to war over. The oil wars in Iraq in the early 1990s and early 2000s were defining military missions for Presidents George H.W. Bush and George W. Bush, but the idea that oil is worth fighting for is actually known as the *Carter Doctrine*. In his 1980 State of the Union speech, President Carter announced that the nation was willing to use military force to protect its interests in the Persian Gulf, at the heart of the Middle East's most active oil-producing region. As an expression of the Carter Doctrine, the US military projects force around the world, deploying the navy to keep sea lanes open so that oil tankers can move freely to their destinations.

Of particular concern are chokepoints. A chokepoint is a geographic narrowing through which commerce flows, making it vulnerable to military blockades or disruption by bad actors.

In 2015, the world consumed nearly 97 million barrels of oil every single day. Of that amount, nearly 59 million barrels were moved by seaborne tankers that traveled through chokepoints.[32] Important chokepoints included the Strait of Hormuz, where ships exit the Persian Gulf (19 million barrels per day, nearly 20 percent of world consumption), the Suez Canal (through which 5.5 million barrels per day of oil flowed to Europe from the Middle East), and the Strait of Malacca, a stretch hundreds of miles long between the Malay peninsula and Indonesia, near Singapore (where 16 million barrels per day traveled en route to Asia). The Strait of Hormuz in particular is considered a risky chokepoint because of its close proximity to Iran and because about one-third of oceangoing oil travels to global markets through this one spot. The concern is that Iran working alone or in concert with other adversaries would seek to close the strait, disrupting global oil supplies.

Because of this concern, the US military keeps an active presence near the Strait of Hormuz, Suez Canal, and oil-producing countries. Forces there are on high alert because tensions can flare up. On June 21, 2004, the longest day of the year, eight members of the UK's Royal Navy and Royal Marines were captured by Iran while conducting river exercises near the strait. Iran accused them of straying into Iranian waters but released them three days later after significant diplomatic pressure. In March 2007, not quite three years later, fifteen Royal Navy personnel were captured with similar claims they were in Iranian waters. The harassment near the Strait of Hormuz continued the following year, this time with American sailors, when Iranian boats sped in front of US Navy ships, dropping large objects into the water.[33] The Americans were concerned the Iranians were dropping mines, but they were in fact empty boxes and were dropped just to be disruptive. The boats switched directions just before they were fired upon. These flare-ups caused oil prices to jump, revealing the jittery nature of global oil markets and international concern about disruptions.

Those same chokepoints also introduced risks from nonstate actors, namely pirates. No longer wooden-legged, eye-patch-wearing swashbucklers from the 1500s who target ships carrying crown jewels, modern pirates attack another form of royal treasure: oil that belongs to the Saudi royal family. One super-tanker can hold 2 million barrels of oil, which at $50 per barrel is a combined cargo worth $100 million. A bounty that large can generate a handsome ransom. In one case in 2009, Somali pirates released the Sirius Star, a Saudi-owned supertanker, for the tidy sum of $3 million after holding it for two months.[34] Pirates attack hundreds of ships each year around the world, and oil ships are tempting targets, receiving dozens of attacks each year.[35] All in all, 470 oil tankers were attacked between 2003 and 2017, with another 133 attacks on natural gas and propane tankers. The frequency of attacks on tankers is significant enough that it has raised global oil prices.[36] Because of this fragility in the global supply chain of oil, many countries are actively interested in reducing their oil imports and achieving energy self-sufficiency if possible.

The United States is one of those countries. In the 1970s, US policymakers erected barriers to stop oil exports on national security grounds. In the 2010s, the United States argued for natural gas exports on national security grounds. In the 1970s, on the heels of the oil crises, the thinking was that exporting US oil to global markets would put oil availability for domestic use at risk because US producers would rather export to the highest bidder than guarantee oil would be available for domestic needs. The export ban was irrelevant because US oil consumption was increasing and production was decreasing, so imports dominated the scene and exports were a whimsical idea. But after the shale revolution—which began outside Fort Worth, Texas, in the

2000s—unlocked domestic supplies, exports became a possibility again and energy geopolitics has been turned on its head.

The 1970s mindset was that the United States needed to prohibit oil exports to get out from under the thumb of Middle East exporters. The 2010s mindset was that the United States must start natural gas exports to get our European allies out from under the thumb of Russian exporters. Consequently, lique-fied natural gas export terminals are under construction in the United States, and Congress—in a rare example of bipartisan action for the era—lifted the prohibition on crude oil exports for the first time since the 1970s. Stateside gas producers can boldly tell Europe that American LNG will solve their energy security problems, too, as it will let them kick their expensive habit of mainlining Russian gas via stainless steel syringes that puncture right into the heart of Western European load centers such as Berlin.

## ENERGY AS A TARGET OF WAR, DARKNESS AS A SIGN OF WAR

Because energy is so critical to society, energy resources such as refineries, dams, or pipelines are often a target of war. Going after the energy supply chain can be just as effective—without the deaths—as attacking the opposing troops. Usually energy is targeted before soldiers are sent in, but retreating troops some-times attack energy infrastructure to keep the advancing troops from gaining access to it. For example, in the 1991 Gulf War, Saddam Hussein's troops in Iraq set oilfields on fire during their retreat so that American forces could not use the oil.

The oil supply chain is the lifeblood of the modern military. The use of petroleum-based fuels has increased 17 percent since the end of World War II, reaching 22 gallons per soldier per day.[37] As a result, during wartime, a lot of effort is spent moving energy around. In 2006, with Operations Enduring Freedom and Iraqi Freedom underway, the US Department of Defense spent $13.6 billion purchasing 110 million barrels of oil and 3.8 billion kilowatt-hours of electricity.[38] Just over $10 billion was for fuel and nearly $3.5 billion for electricity. Because of increasing consumption and higher oil prices, the military's price tag for fuels in 2008 was $20 billion.[39]

In total, fuel logistics is responsible for about 70 percent of the tonnage the military must move into position for battle.[40] The cost is extraordinary. Mid-air refueling for the US Air Force costs $42 per gallon. While pricey, that's eclipsed by the $400 per gallon needed to get fuels to the front line. There are about 60,000 uniformed members in the army alone—20,000 active duty and 40,000 reserve personnel—just to move fuels worldwide. Making matters worse, with the rise of new equipment such as unmanned aerial vehicles (e.g., drones), there has been a proliferation of fuels. In 1990, there were six types of fuel in use on the battlefield. By 2005 there were ten and by 2010 there were thirteen, each one needing to be delivered for its use in theater.[41]

The extended supply chain to deliver those fuels creates inviting targets for insurgents. Unfortunately, that leads to more risks, as the delivery trucks are less capable of protecting themselves. Fuel convoys are easy targets for roadside bombs, so in recent wars, more fuel consumption has directly correlated with more deaths in theater.[42] There were over 150 deaths involving fuel convoys in Iraq and Afghanistan in 2007 alone.[43] The army found that switching to solar power for forward operating bases (FOBs) reduced the need for fuel, thereby directly saving lives.[44]

Pipelines are another part of the fuel supply chain that make for tempting targets. During a tussle between Russia and Georgia in 2008, Russia used the pipelines as a way to extract leverage from consuming nations and to get support for the NordStream 2, a proposed pipeline along a more northern route.[45] Newspapers reported that multinational energy companies were on alert as they watched the world's second-longest pipeline, which moves 1 million barrels per day across Georgia, suffer the onslaught of dozens of missiles from Russia. Thankfully, the missiles missed. Whether these strikes were intended to impact Georgia or the Western customers for the crude was not clear. Oil prices were very high at the time, which gave Russia—a country blessed with abundant oil and gas—an influx of money and confidence. They could issue muscular missives from their arsenal of petro-militarism. What's remarkable is that the missiles just missed the pipelines. Expert observers noted that Russia's navigation systems are better than that; thus, they likely missed intentionally. It's as if Russia was sending a signal that pipelines through Georgia were unreliable and could be taken off-line remotely on short notice by their missiles. It might have been a way to recruit business for more expensive, northerly Russian pipelines, as the country would be unlikely to destroy its own pipelines. The European Union, which is proximate and at greatest risk of the fallout from energy conflict between Russia and its neighbors, sat silently. That silence acknowledged European dependence on Russia for oil and gas. A decade later, at a famous summit with Vladimir Putin in Helsinki, President Trump recommended the NordStream 2 pipeline as a way for Europe to improve its energy security, finally helping Russia achieve that goal.

Intentionally knocking out the power is also a typical war tactic because it hurts an opponent's ability to execute their warplan and undermines the morale of the citizens. However, destroying the energy infrastructure inhibits the possibility of postwar recovery in the vanquished territory, which can prolong instability and inhibit peace. To navigate this trade-off, during the war in the Balkans in the 1990s, the United States deployed a new type of weapon: graphite bombs. Also known as "blackout bombs," they cause blackouts but without destroying the grids.[46] The bombs spread out a cloud of carbon filaments a few hundredths of an inch thick that caused short circuits as they floated down and landed on uninsulated electrical transmission and distribution wires. This is useful because as soon as operations are over, the grid doesn't have to be rebuilt. In an intensive bombing campaign in May 1999, NATO planes cut the power in Belgrade, the capital of Serbia. At one point, the lights went out over 70 percent of the country.[47] Serbia was able to clean up the grid and restore power within twenty-four hours, so the North Atlantic Treaty Organization (NATO) did the same thing again a few days later. NATO spokesman Jamie Shea said, "We can turn the power off whenever we need to and whenever we want to," which had a great psychological effect on the residents.[48]

And potential attacks on the grid become more advanced as our technology does: for example, the possibility of high-altitude nuclear attacks that would create an electro-magnetic pulse (EMP). EMPs can occur from natural causes such as lightning, or from purpose-built weapons. Similar to what happens during electrical storms, the pulse of energy creates surges that damage equipment such as transformers or substations, which knocks out portions of the grid. These power outages disrupt life and interfere with the operations of domestic military bases that use the local grid for their power. They could also turn off refineries, which produce the jet fuels needed on the battlefields. This risk is sufficiently real that its possibility has been acknowledged in military planning documents.[49] The biggest challenge isn't just the disruption itself, but the possibility that it would endure for an extended period of time, beyond the capability of backup systems. In particular, for some large equipment in the grid, such as high-voltage transformers, there is only one manufacturer in the world, located in Canada. The fears about EMP attacks grip the imagination because it is hard to contemplate a modern society without electricity. Utilities are sufficiently concerned about the risks that they are actively studying the scenario. Their assessments indicate that a nuclear bomb 100 times more powerful than what was dropped on Hiroshima could cause regional outages. Fortunately, however, they also concluded that none of the scenarios they evaluated would cause a nationwide grid collapse.[50]

When the grid goes out, it is hard to bring it back on. In fact, the process of bringing the grid to life after a major outage is called a "black start." Most power plants cannot operate unless the power is already on. To bring a power plant on, the grid needs to be supplying about 10 percent of its rated power; otherwise the plant's equipment cannot synchronize. Because of this challenge, hydroelectric power plants are often used for black starts because gravity still works even if the grid is down. And gravity is still free. But not all regions have abundant hydroelectric power available to prop up the grid. In early May 2018, my research group at the University of Texas at Austin went through black start training with the Texas grid operator. Texas has very little hydroelectric capability because of scarce water resources. The instructor for the workshop told my students, "I keep a twenty-four-day supply of water in my garage. If the guy in charge of black start training does that, that should tell you something."[51]

## THE DOD MIGHT SOLVE OUR ENERGY PROBLEM

Although many look to the US Department of Energy to take the lead on energy issues, it is one of the ironies of the federal government that, contrary to its name, the Department of Energy historically has been a part of the national security apparatus; it's a weapons agency that happens to dabble in energy on the side. The flip side is that the Department of Defense (DOD) is much more involved with energy issues than expected, primarily because no organization in the world faces more complex energy-operating issues.[52] The department is the world's largest energy consumer, a guarantor of energy trade, a victim (and beneficiary) of energy-related military tactics, and a strategic protector of US interests in energy-rich areas of the world.

All these different uses for energy as a weapon, transporter for weapons, or guidance system for weapons means that the military consumes vast quantities of energy. In total, the DOD is the world's largest single energy customer, consuming more energy than entire countries like Bulgaria, Denmark, Israel, and New Zealand. The DOD consumes 5 billion gallons of liquid fuels a year. Of that, 3.6 billion gallons is jet fuel alone, mostly for the air force. In all, the DOD consumes about 700 trillion BTUs, which is about 75 percent of all US government consumption and nearly 1 percent of total energy consumption across the United States for all purposes. By comparison, that is sixteen times as much energy as what the US Postal Service needs to operate the largest civilian fleet of vehicles in the world, topping out at more than 230,000 vehicles that drive more than 1.5 billion miles each year.[53] During the height of the Cold War in the 1970s and 1980s, when the military maintained a much larger fleet of strategic aircraft that were airborne at any given moment, the consumption was nearly twice as high as it is today. The good news is that energy consumption by the US military has dropped considerably, partly because of the end of the Cold War and partly because of new efficiencies that have been developed. Because its fuel needs are so specific, the DOD is a microcosm of the broader societal energy challenges: ensuring that energy is reliable, affordable, and sustainable.

Like the US Postal Service, the US military has hundreds of thousands of ground vehicles, ships, and planes. The US Navy also needs fuel for its ships, so they turned to nuclear power for many submarines and the aircraft carriers. Ships are remarkably efficient, so their fuel requirements are small compared to tanks and planes. In the end, aircraft are the biggest driver of energy consumption by the military.

Compounding the challenge, the energy requirements for jet fuels are very specific. When it comes to fueling planes or tanks, the US military needs higher-performing fuels than the average Joe filling up his truck at the corner gas station. The air force in particular needs fuels with really high energy density so that they can keep their fuel tanks light and compact, and they have to be safe to store, handle, and use. These fuels must also have convenient boiling and freezing points to ensure they work in the scorching heat on a tarmac in the Middle East in the summer or the chilled temperatures of Alaska in the winter or at high altitudes. These requirements make electricity stored in batteries and ethanol (both of which have energy densities much lower than petroleum-based jet fuels) impractical for most heavy-duty or long-haul aviation applications. It is hard to use electricity for aircraft because conventional batteries are heavy. Biofuels do not have the same energy density or freezing and boiling points. Consequently, the Department of Defense would have difficulty using conventional, off-the-shelf "alternative" energy solutions from the ground-based transportation sector to meet its needs.

The US military experimented with nuclear aircraft starting as early as 1946 with the Nuclear Energy for the Propulsion of Aircraft (NEPA) project. They thought that nuclear reactions (rather than burning jet fuel) would generate the high-temperature and high-pressure gases needed to accelerate out the nozzle to propel the plane. But protecting the crew from radiation onboard was a vexing problem, and the idea of nuclear-powered aircraft flying overhead made citizens nervous. The nuclear aircraft programs, which existed in multiple forms and program names and yielded a few different test engines, were ultimately stopped during the Kennedy administration. In the end, liquid fuels provide the best balance of performance characteristics and are the key driver of military energy consumption.

Because energy use has a critical impact on the Department of Defense's bottom line, the department has a much greater financial incentive to find ways to reduce its consumption than the Department of Energy or any other government agency. And because agency budgets are often a zero-sum game, if the Department of Defense can reduce its bill for fuels, then it might have more money available for other purposes. As a Defense Science Board study noted in light of the high costs of delivering fuel to the front lines for military use, "It is unlikely that energy efficiency has a higher value to any other organization in the country, possibly the world."[54] So the Department of Defense is also increasingly becoming a leader in energy research and development, and in implementing new energy technologies and using alternative fuels. Therefore, the department is a top contender for finding a low-carbon, domestic source of energy that is compatible with existing infrastructure and satisfies the country's energy needs just as well as traditional forms of energy. If the Department of Defense can do that, then it will not only solve the United States' energy woes, it will also go a long way toward solving the world's energy challenges.

Perhaps the most important tools at the Department of Defense's disposal, however, are its significant research budgets, including those for the Defense Advanced Research Projects Agency (DARPA) and the research arms of the different branches, such as the Office of Naval Research (ONR) and Air Force Office of Scientific Research (AFOSR). The total amount of R&D money related to energy in the federal government's entire research portfolio usually hovers around $3 billion per year. That's only a fraction of the more than $30 billion we spend on health-related research and miniscule compared to the $80 billion we spend on defense-related research. Based on the sheer size of its investments, the Department of Defense has more capacity to conduct effective energy research than all other government agencies combined.

Keeping the hydrocarbons flowing is a top strategic priority because energy is such a critical economic and political concern for the United States and its allies. The ironic result is that the US military expends a significant portion of its energy stabilizing or protecting energy-rich parts of the world. It's hard to put a price on this total effort, but different analyses estimated that the United States spends somewhere between $29 billion and $143 billion every year just to protect the supply and transit of oil, not including the cost of wars in Afghanistan and Iraq. These expenses mean Americans are paying anywhere from about $9 to $44 of additional hidden military costs for each barrel of petroleum we import from countries other than Canada and Mexico. When I was part of a national security analysis team at the RAND Corporation, I remember hearing on more than one occasion while meeting with senior officials at the Pentagon how the fall of the House of Saud or an attack on their major refinery would be a global security problem.

All of this energy-inspired intervention worldwide gives many planners in the Pentagon a grave philosophical headache. Thankfully, the Department of Defense has some critical policy resources at its disposal to solve its own energy problem.

It has firsthand experience producing, distributing, and using energy, which comes in handy when developing alternative fuels. In addition, DOD has a unique ability to enter long-term purchasing agreements for power (twenty years or more) and fuels (up to five years), which gives energy developers the revenue stream they need to secure financing to bring large-scale projects online. Other governmental agencies have difficulty entering into long-term contracts.

As a result, the US Air Force set ambitious goals to use alternatives to petroleum, which could include biofuels or liquid fuels made from coal. They also actively flight-rate their planes with the new fuels as new fuels emerge. While biofuels are attractive to policymakers because the plants grow by taking $CO_2$ out of the atmosphere, the ethanol that is typically produced is a poor jet fuel because it has low energy density. But jet fuels similar to biodiesel can be made from a few different processes using a range of feedstocks. One of these—algae—is particularly interesting because it is potentially much more productive per acre per year than traditional biofeedstocks such as corn and soy. And because farmers can grow algae with low-quality water (including saltwater in some cases) on nonarable land, algal biofuels can avoid competing with agriculture.

Researchers have known for decades that algae are a source of fuel, but developing it has been plagued by prohibitive cost and scale hurdles. That's where DARPA—the R&D arm of the Pentagon famous for creating the Internet—comes in. In the late 2000s, DARPA invested more than $50 million for a crash-course, twenty-four-month program to generate thousands of gallons of algae-based jet fuels at a price competitive with gasoline. As of 2018, those investments had not yielded commercially available quantities of algal-derived biofuels, but they did help trigger some significant follow-on investments from Exxon-Mobil, the world's largest publicly traded oil company.

In addition, the department has embraced renewable energy. In 2005, the US Air Force became the largest purchaser of renewable energy in the Environmental Protection Agency's (EPA's) Green Power Partnership (a voluntary program that supported the procurement of green power). While the air force is no longer in first place, it installed what at the time was the nation's largest solar photovoltaic array, with a 15-megawatt capacity, at Nellis Air Force Base in Las Vegas, Nevada, the same AFB where I watched drones fly over the Middle East. And a few bases, including Dyess in Texas and Fairchild in Washington, receive 100 percent of their energy from renewable resources, while large bases such as Fort Carson in Colorado push to be net zero for energy and water.

The military can also provide intellectual leadership. While the national political debate about climate change rages on, the Quadrennial Defense Review (QDR), a strategic defense plan generated every four years, made no hesitation in its declaration that climate change is a threat to national security.[55] The QDR from 2014 notes that climate change poses profound strategic challenges for the United States and increases the likelihood of military action because of the need for intervention to deal with the effects of violent storms, drought, mass migration, pandemics, toppled governments, enhanced terrorist movements, or destabilized regions. As reflected by the QDR's conclusions, the US military's view is that mitigating climate change will reduce the need for those interventions and improve the national security situation. A 2008 Defense Science Board study recommended that the DOD should invest in low-carbon fuels to send a signal that the United States would collaborate in an international effort to reduce $CO_2$ emissions, but also to help mitigate climate change as an exacerbation of national security threats. That recommendation is one of the reasons the military has pushed for biofuels or other low-carbon options.

In the end, the Department of Defense has a large responsibility to help the world avoid energy crises—and does so by stabilizing energy-rich regions and preventing supply cutoffs. But it comes at a huge cost in energy, money, and personnel. This complex relationship with energy, combined with the department's purchasing power, gives it unique motivation, insights, and capability to solve the world energy problem. And, in fact, it's already doing more than most people realize. The United States has spent decades building the most capable military in the history of the world, and energy just might be one more area where it achieves success.

## NUCLEAR SECURITY

Mention "nuclear security" and the first thing that comes to mind will be warheads. But nuclear power is relevant to the story of energy in other ways.

Nuclear power is very useful as a fuel source for navies. Because of the supreme energy density of nuclear fuels compared with fossil fuels, large vessels like aircraft carriers use nuclear reactors as a way to save a lot of space. But because of the technical complexities, cost, and international treaties, only a handful of countries have nuclear-powered aircraft carriers. Nuclear power is also particularly useful for submarines because it allows them to remain underwater for much longer periods of time than submarines powered by batteries and diesel. Diesel submarines require oxygen to operate and therefore must stay close to the surface and submerge for limited times. And batteries have limited duration before they must be charged again.

But nuclear power in the electric sector is a mixed blessing. Since climate change is a national security threat, using low-carbon nuclear power domestically to avoid emissions of greenhouse gases has an environmental and security benefit.

And promoting nuclear power in poorer parts of the world might have the benefit of reducing poverty, thereby reducing the threats of terrorism or civil unrest. But these nuclear fuels can be enriched to levels that are suitable for weapons, opening up concerns about weapons proliferation. It is this risk that makes the global community so wary of Iran's and North Korea's desires to use nuclear power for electricity generation. It also explains why developed countries should keep their nuclear programs alive: to improve global nuclear security.[56]

Nuclear power faces many hurdles around the world. Power plants under construction are facing serious delays, halts, and cost overruns. And several existing nuclear power plants are at risk of shutting down due to high costs, unfavorable public sentiment, or lack of competitiveness with cheap natural gas or wind and solar energy. What many don't realize is that these closures pose long-term risks to national security by discouraging the development of our most important anti-proliferation asset: a bunch of smart nuclear scientists and engineers.

This is the irony of nuclear power: while many worry that the prominence of nuclear materials increases the risks of weapons proliferation, the opposite is also a problem. The loss of expertise from a declining domestic nuclear workforce makes it hard for the United States to provide experts to conduct the inspections that help keep the world safe from nuclear materials.

The Defense Threat Reduction Agency (DTRA), the US agency responsible for addressing these risks directly, employs 2,000 people to tackle chemical, biological, radiological, and nuclear weapons. Hundreds work on the nuclear mission alone. Another 2,500 people work at the International Atomic Energy Agency (IAEA), a multinational organization created for the sole purpose of ensuring peaceful uses of nuclear energy. The IAEA is tasked with conducting regular inspections of civil nuclear facilities and auditing the flow of nuclear materials and experts.

Quite simply, it is in the US national interest to maintain the expertise needed to staff the DTRA, while also contributing to the international agencies committed to keeping the world safe from nuclear weapons. In the United States more than 50,000 people are currently employed making nuclear fuels or at the power plants that use them. If the nuclear industry is allowed to wither, we might not have the homegrown talent to help manage the risks. That void would likely be filled by the French or Japanese, which might be fine since they are important allies, but it also makes us dependent on their experts.

Bailing out decades-old power plants with government handouts or subsidies seems like a step backward, but putting a price on carbon would harness the efficiency of markets while allowing nuclear power to compete because of its low-carbon footprint. There are already strong economic, logistical, and environmental reasons to keep nuclear a part of the national fuel mix. Enhancing our national security makes the argument even more compelling. It also lets us focus on other, less obvious security threats.

## POT, SOLAR POWER, AND TAXI DRIVERS: ENERGY SECURITY AUSTIN-STYLE

Energy security means different things to different people. Energy security in the United States means reducing imports of oil from the Middle East. In Europe, energy security means reducing imports of natural gas from Russia. But in Austin, Texas, it means something else entirely.

As discussed earlier, we use oil for transportation in the form of gasoline and diesel fuels for our cars and trucks. Other than in Hawaii and a few diesel generators here and there for backup power, oil is not used for making electricity in the United States. Despite that, it is a common claim by politicians or enthusiasts that they want to install more solar panels or wind turbines to improve our energy security. While wind and solar might displace coal use in the power sector, they don't displace our oil imports. That means wind and solar don't improve the energy security challenges associated with importing oil from countries with different foreign policy goals. This nuanced point isn't obvious to most observers and is an important part of the classes I teach at the University of Texas to some of the world's smartest graduate students.

One day, while riding in a taxi to the airport in Austin, I struck up conversation with the driver, which is a typical habit for me. When I told him I study energy for a living, he was very excited. "I love energy," he said, joyously. "So do I," I responded, sharing his enthusiasm. He then went on to say that he especially likes solar energy because it improves our energy security situation.

With a slight eye-roll to myself in the backseat, I was about to start correcting his mistake by explaining in snotty professor-speak how solar energy does not reduce oil imports and so it doesn't improve our energy security. Before I could start my lecture, he continued: "Yeah, if we all had solar panels, then we could grow our own marijuana at home, which means we would not have to import it from warlords and violent cartels, and that would improve our national security."

My jaw dropped. Indeed, he was correct. I hadn't thought of that. Solar energy does improve national security. I learn something new every day.

## LOOKING FORWARD

Just as energy has reshaped warfare in the past, we can expect it will continue to do so. The question that lingers is whether energy has made us a more secure or a less secure society, and that question has no easy answers. Advanced defense systems powered by electricity or other modern fuels make us safer. But our dependence on unstable regions for our energy supplies does not. Further, climate change influenced by rampant energy consumption destabilizes regions and raises the risk of conflict. As it is, Defense Science Board studies and Quadrennial Defense Review reports prepared for and by the Pentagon note the importance of taking climate change seriously.[57] The Defense Science Board called for the DOD to be a leader in pushing for low-carbon energy solutions. The 2014 QDR noted that the military would be stressed by climate change since it "may increase the frequency, scale, and complexity of future missions, as well as undermine the capacity of our domestic installations to support training activities." It went on to call for "investments in energy technology [that] will make for a stronger and more effective fighting force." From the perspective of the Pentagon, getting the energy needs of the military aligned with the low-carbon needs of society as a whole is a critical priority to reduce the risk of conflict.

Morgan Bazilian, executive director of the Payne Institute at the Colorado School of Mines and formerly at the United Nations and World Bank, noted, "Good planning, energy infrastructure, and well-managed service delivery can be a basis for peace and economic development. The reverse is also true."[58] There is no doubt that energy shortages can and will continue to be a cause of conflict. But designing new energy systems that increase availability of water and energy resources might be a pathway to peace.

# THE FUTURE OF ENERGY

## TRADE-OFFS

Life is, in many ways, about compromises. And as I hope this book has shown, life as we know it is also about energy. So energy, too, is often a matter of compromises. Though energy is all around us, it is still complex and hard to understand. Every fuel and technology option has benefits and risks. It's folly—and typical in the contemporary political climate—to advocate for or against one fuel or another without admitting the contributions of or problems with each. Coal backers point to its domestic abundance and low cost as a sign of its security and economic benefits, neatly sidestepping its environmental impacts. Backers of renewables point to the cleanliness of wind or solar, diminishing their expansive land requirements or variability.

There are even trade-offs relevant to the quantity of energy we consume. For example, consuming energy can make us more wealthy, but being more wealthy makes us consume energy, and consuming too much of the wrong kind of energy can actually make us *less* wealthy, again because of the accumulating public health and ecosystem costs from environmental damage that makes people sick or hurts other industries such as tourism and agriculture.

Today coal is considered a threat to the forests because of mountaintop removal mining and acid rain, but in the 1800s coal saved the forests because it slowed deforestation. Today oil is considered a threat to whales because of impacts from offshore production and ocean shipping, but in the 1800s oil saved the whales because of kerosene's role as an alternative to whale oil as an illuminant. Those trade-offs make it harder to develop simple solutions to a complex problem. That fact, combined with the evolving nature of our challenges, means that yesterday's solution might be tomorrow's problem.

Looking forward, which of today's solutions will be tomorrow's problems?

In the modern world, everyone uses energy with a mixture of elation and guilt: it's the energy-lover's dilemma.[1] How do we get all the benefits we enjoy from consuming energy without the downside impacts? Because each fuel and technology has its upside benefits and downside risks, energy managed the right way offers tremendous benefits to humanity. It makes us wealthy, healthy, and free. But energy done the wrong way introduces vulnerabilities and catastrophes, reducing our security, increasing economic inequality, and poisoning the environment. Responsibly improving access to the creative potential of energy is the grand challenge of the twenty-first century.

## THE TRENDS AND TRANSITION

As we tackle this grand challenge, we must keep in mind that there are six demographic trends, three technology trends, and one overarching environmental trend affecting the energy system. I told you it was complicated.

The six overarching demographic trends—population growth, economic growth, urbanization, motorization, industrialization, and electrification—drive our energy demand. Because of population growth, there are more people who want access to energy, food, and water. Each person requires food and water to stay alive and needs energy in some form or another to gain access to those resources. On top of the population growth is economic growth. It is a nearly universal trend that rich people consume more food, water, and energy per person than poor people.

Industrialization makes us richer. Access to energy gives us the opportunity to produce goods and services that let us accumulate wealth. So consuming energy (for industrialization) can make us rich, which causes us to consume even more energy.

As people become wealthy, they move into cities, buy cars, and electrify as many activities as possible. Electricity is the preferred form of energy for rich people because of its convenience and cleanliness inside our homes and businesses. Cars give us the freedom to travel independently but also typically require fuels like gasoline or diesel. Just as cars were symbolic of American freedom in the postwar era of the 1950s—the heyday of American automotive manufacturing might—the same happens elsewhere. Economically developing countries see rapid adoption of automobiles as a status symbol.

As societies get richer, their citizens switch to diets with more protein, moving from tubers, seeds, and nuts to animal proteins. Animal proteins are notoriously water- and energy-intensive because of the fertilizers and irrigation needed to grow the feed.

These six trends combined are driving this shift in the amount and type of energy we consume. Instead of using energy just for heating and cooking, which can be done with solid fuels like straw, cow dung, and firewood, we are using energy for cars, factories, and power plants. That means we're adding petroleum, natural gas, and electricity (made from a variety of fuels, including coal, nuclear, wind, solar, and water) to primitive fuels.

On top of the six demographic trends are three technological trends. First, society is becoming more efficient. We use less energy to achieve the same purpose today for our products and services than we required decades ago. Whether it's lighting a room, driving a car, or shaping metal, we can do it with less energy than before.

Second, everything is becoming more information-intensive. Because of ubiquitous sensing and cheap computing, we have more data available than ever thought reasonable, even as recently as the birth of the internet. This information is one cause for society's increasing energy efficiency—we can fine-tune our processes to optimize energy usage and reduce waste—and it is changing the way we interact with each other, appliances, and buildings.

The third technological trend is the rise of decentralization. Rather than making electricity at large power plants far away, it is becoming common to make electricity on-site with rooftop solar panels. Instead of relying on far-flung factories, the rise of 3D printing allows for on-site manufacturing. That means mass customization instead of mass production. In the place of mass transit systems on fixed schedules along inflexible routes, the rise of micro transit and mobility services such as Uber and Lyft brings with it point-to-point transportation that is faster and cheaper and gives people what they want. Ultimately, decentralized access to energy is synonymous with democratized access to energy. In his history of the industrialization of light, Wolfgang Schivelbusch observed, "The concentration and centralization of energy in high-capacity power stations corresponded to the concentration of economic power in the big banks."[2] The implication is that decentralization is liberating and helps us overcome income inequality, while autocratic control of energy is confining. So perhaps that means we can be hopeful the movement to distributed forms of energy (rather than large power plants) will lead to a better distribution of wealth in society.

These technological trends mean our energy systems are becoming smarter, faster, and cheaper. All of which is good news because the combination will help us manage the trade-offs of energy's upsides and downsides while pursuing the final contextual trend. In fact, the future might see a convergence of information, energy, and our daily lives. My good friend and mentor Roger Duncan, the former CEO of Austin Energy, who was listed in a 2005 issue of *Businessweek* alongside Tony Blair and Arnold Schwarzenegger as one of the world's most important people because of his work on reducing power sector emissions, refers to this future as one in which we have sentient-appearing machines and buildings.

This overarching trend is the overall combined market and policy outcomes of decarbonization—reducing society's carbon emissions through shifts in agriculture, land use, and energy. Reducing the carbon intensity of our fuels—switching from coal to petroleum to natural gas, for example—has been underway for over a century. But our energy consumption has grown significantly in the meantime. Today, when our energy consumption is growing more slowly and our fuels are getting cleaner, our emissions in the United States are actually dropping despite ongoing population growth. Hopefully, this trend will hold in the developed world and gain traction in developing countries like India and China.

In many ways, decarbonization in parallel with increasing energy access is the defining challenge of the twenty-first century, the way industrialization and liberation from slavery and imperialism defined the nineteenth century and defeating fascism defined the twentieth century. Decarbonization is critical to avoid the worst effects of climate change. But since there are still more than a billion people without access to electricity, piped water, or sanitation, we also need to focus on energy access. Because of what it does to extend and improve lives, making energy accessible is as important as making it clean and sustainable.

We must embrace the wonderful things brought to us by energy; we can't quit using energy or deny others access as a solution for decarbonization. While the decarbonization trend in the United States and Europe might feel unstoppable by inertia alone at this point, there are still many pitfalls along the way to decarbonizing society.

## WE NEED A NEW WAY OF THINKING

Energy is intertwined with society in so many obvious and non-obvious ways. If there is one lesson we should learn from our energy challenges, it is that there are no universal, immediate solutions. Worldwide energy use is a complex system with many parts, and improving it will take sustained efforts and a deep historical understanding of energy.

Rick Smalley, who gave the world his list of ten grand challenges, concluded that to solve our energy problem we need to do three related things: (1) inspire the next generation of scientists and engineers, (2) develop energy supplies that won't run out, and (3) solve global climate change.[3] These are daunting indeed but also invite new opportunities for investment, innovation, and advancement.

The way to solve this conundrum is not by hashing out old clichés of fossil fuels versus renewables or other tired battles. The same thinking that got us into these problems—drill more, pave more, consume more—will not get us out of them. Tired techno-enthusiasm that just says we can use smarter gadgets won't be enough, either. And reducing our consumption to the levels of energy poverty won't be very satisfying or humane. We need cleaner forms of energy, such as nuclear, wind, or solar, or we need to clean up conventional forms with carbon capture and scrubbers so that we can maintain and expand energy access without scorching the planet. If we act too slowly, then we might need geo-engineering to clean up our mess after the fact, to scrub $CO_2$ from the atmosphere. We also need to transition from a business model oriented to energy production to one that prioritizes energy services. That way we get what we want—the energy services, such as heating, lighting or mobility—without the profit motive to simply produce more energy. Ultimately, we need some combination of new production, increased energy access, smarter solutions, and a cultural emphasis on efficiency and conservation. Only then can we simultaneously mitigate energy's problems while bringing its benefits to more people worldwide.

We need to approach energy with a whole new way of thinking. By looking at energy with a global, long-term view, we will enable innovations that reduce waste, bring forth efficiency, and mitigate environmental impact.

The long-term view has been missing from typical American energy decision-making. Often our energy options are posed as a choice between the economic path (the cheaper one) or the environmental path (the cleaner one). This is a false choice because, clearly, Americans want something that is both cheap and clean. And it's an artifact of short-term thinking. In the near term, the environmental option—better windows, higher-performing air conditioners, more efficient automobiles—often costs more money up front. But in the long run, the environmental solution and the economic solution are the same. More expensive, efficient, cleaner items usually reduce operating costs.

At the same time, pollution inhibits economic growth because sick people do not work as much and degraded ecosystems are less productive. In all, the cumulative environmental impact of dirty options becomes very costly over time. If we shift our time horizon from the immediate future to the coming decades and include the costs of failing to protect the environment, then the distinctions disappear, making our false choice between the economy and environment irrelevant.

Our thinking also needs to be more global in nature. If we all just focus on our own backyards, then we ignore the pollution our choices make in other people's backyards. Pushing our power plant into someone else's watershed does not reduce pollution; it shifts pollution, often to those who are least empowered to do anything about it. But when we recognize that our resources—and pollution—flow with global reach, then our decisions gain moral clarity. One way to think about it is as if the entire earth is our backyard, in which case we would feel an imperative to reduce pollution wherever it occurred.

Global climate change is the ultimate example for why we need long-term thinking and a global perspective. The decisions we make today with our resource usage will affect kids on the other side of the world who haven't been born yet. Near-term, local thinking will not begin to solve this problem. There's no such thing as other people's kids. All the world's kids are everyone's responsibility. Either we care about them or we don't. Making sure they will have access to energy in a habitable environment is our most pressing obligation.

Just as the keystone is the one critical piece that keeps Roman arches standing thousands of years later despite the strain they endured while expanding the reach of Roman culture, energy is the critical piece that will keep modern society prosperous for thousands of years while expanding the franchise of liberty.

Ultimately, to solve our energy challenges, the world needs the next generation of innovators. We need them now. Thankfully, our younger generation of future leaders is creative and speaks with moral clarity. From the position of having taught over 1,000 college students in less than a decade in a traditional classroom setting plus another 5,000 students from around the world through an online course I offered, I have confidence they will step up to guide us through this jungle.

Let me point to my daughter Evelyn Webber for two examples from when she was eight years old.

One night when she was sick and couldn't sleep, she joined me in the living room where I was watching a documentary about corn ethanol as a substitute for gasoline. It was hours past her bedtime and she was not feeling well and was very silent, so I thought she was sleeping or lost in her own thoughts. But it turns out she was paying close attention and that she was outraged. When it ended, to my surprise, she suddenly bolted upright. She grabbed her pad of lined paper and a red pen and then with her best cursive in a hurried rage wrote a two-page essay without any hesitation titled "Why We Can't Use Corn!"[4] Her opening line was "We should not use corn to make oil because we eat it!" In her first sentence she revealed her intuitive understanding of the food-versus-fuel dilemma posed by modern biofuels. She then continued, explaining that instead of using new resources, we should use trash and other forms of pollution for our energy.

One other time when she was eight years old, I was driving her to a soccer game, more than 25 miles away from home.[5] We were stuck in traffic with a seemingly endless stream of cars, when she said, "If everyone keeps driving cars so much, then the world will run out of gas, and that will be great because then we will have to ride bikes, which will be fun." She's right. Biking would be a lot more fun than sitting in traffic, trapped inside a metal cage, isolated from the people and nature around us.

It's so easy to get stuck in our patterns and assume that's the way it needs to be without stepping back to question whether that's the way we want it. For many people, the idea of getting away from high levels of oil consumption and individual car ownership is scary. But from my daughter's perspective, it would be liberating. Her reactions and observations revealed to me that in many ways children reflexively draw these connections between our quality of life, the Earth's health, and the need for mindful stewardship of our resources. Over time, that instinct gets muddled for reasons I can't explain. But returning to that childlike wonder and releasing ourselves from the dogmas attached to one form of energy or another opens up the possibility for identifying whole new solution sets.

At its core, energy is nothing more and nothing less than a way to change the order of the world. Energy is magical. After all, it made the world as we know it, and it will probably remake the world many times over. So there's good reason to believe that energy will provide the answers we need, as long as we are willing to understand it. It might seem silly or an abdication of our responsibility to look to a younger generation to solve our problems. Especially considering that everyone who has contributed to energy's current state of affairs was also once a child. But solving energy's problems requires imagination—lots of imagination, from lots of people. Energy is the stuff dreams are made of. A little childlike wonder to give us an appreciation of energy and an open mind about how to solve the energy challenge might be the right combination to make progress.

To help foster our children's ability to solve our resource challenges, we need a new approach to energy education.[6] A scientific study noted that people fall for logical fallacies when believing misinformation and rejecting the scientific evidence of climate science.[7] That means we need an emphasis on critical thinking so that we can more effectively distinguish fact from fiction. Science, technology, engineering, and mathematics (STEM) education teaches analytical thinking, but we also need liberal arts, which gives us critical thinking. We also need fine arts, which gives us creative thinking that we can use to seek new solutions. A common joke goes like this: the *earth* without *art* is just "eh." In that spirit, we don't just need STEM, we need STEAM, where the A could represent the "arts" from fine arts and liberal arts.

Unlike many other aspects of society that have centralized authority, energy has 325 million decision-makers in the United States alone and more than 7 billion decision-makers worldwide. Each of us is making a choice with every appliance we switch on, the houses we buy, and how we live our lives. There are many fuels and technologies that we can implement. But fancy gadgets and clean fuel substitutes are not enough. We also need to change our mindsets and our behaviors. Conservation and efficiency are two of the few solutions that work at every time scale and spatial scale. We can conserve today at the scale of an individual by turning off a light or adjusting a thermostat. We can also conserve at the scale of a region for centuries by redesigning our cities to be walkable and compact.

Adopting a cleaner suite of options while increasing energy access and letting go of our dirtier past is our critical path forward. But just denying access to those in energy poverty or telling rich people in developed countries they will have to go without will hit its own roadblocks. By focusing on energy's magical capabilities, we can solve this challenge without hating energy or making it the enemy. Change is good, so we should go for it. But change is slow, so we'd better get started. We all have a stake in the outcomes and there is much work to be done, so we need all hands on deck, as everyone has a role to play.

# Acknowledgments

I could not have written a book with this breadth without learning a lot from my students and other researchers who collaborated with me at the University of Texas at Austin and elsewhere. I am very grateful for their work to advance the state of scientific understanding. There are too many people to name, but a few of them are called out with more specificity below.

I also could not have written this book without the financial support to do so. I want to thank Doron Weber and the Alfred P. Sloan Foundation for their support via a book grant, which gave me the resources and time I needed to draft the book despite my various other professional commitments. Research support and other financial contributions from the Sloan Foundation, Cynthia and George Mitchell Foundation, Texas State Energy Conservation Office, US Department of Energy, National Science Foundation, Itron, and many other individuals, foundations, and corporate sponsors helped generate some of the new facts and findings that underpin this book. A hallmark of an advanced society is its willingness to invest in the future, so those who are willing to sponsor forward-looking research are part of a rarefied and important group. I thank them for their confidence in my students and their willingness to give them a chance to succeed. This book is just one manifestation of those sponsors' generosity.

Eric Henney and T. J. Kelleher, my editors at Basic Books, and project editor Stephanie Summerhays and copyeditor Beth Partin were excellent collaborators and gave me many useful suggestions for revision. Melissa Chinchillo at Fletcher and Company represented me well. Many of the chapters contained sections that were originally fleshed out for magazine articles. I want to thank Mark Fischetti at *Scientific American*, Megan Sever at *Earth* magazine, and Jeffrey Winters at ASME's *Mechanical Engineering* magazine. They are all very patient editors and have worked with me on multiple occasions to help me turn my various ideas about energy into words that other people can understand.

I have received important coaching along the way, which helped make this book possible. The Presidential Leadership Scholars (PLS) program, which is a bipartisan leadership training program organized by the George W. Bush Center in Dallas, Texas, and the William J. Clinton Center in Little Rock, Arkansas, gave me the opportunity to hone my leadership and thinking skills over a five-month intensive period in 2018 that included guided self-reflection and collaboration with emerging leaders nationwide. That the presidents and their staffs take the time to meet with people like me is testament to their belief in our potential to solve problems collaboratively. I learned much from them. In addition, I have worked closely with Johnnie Johnson of World Class Coaches, which helped me stay focused on my vision and mission. I would like to thank Johnnie and the PLS programs for giving me the boost I needed.

Writing this book required me to expand my knowledge beyond the typical scope of my energy teaching and research. For my process of self-guided learning, some books were particularly instructive to me. Vaclav Smil is a prolific writer whose holistic view aligns well with my thinking and has taught me a lot. His books are standard-bearers and he continues to produce time-less work that is a useful starting point for energy researchers and enthusiasts. My friend Martin Doyle's book *The Source* was a fantastic resource for the role of rivers in America. And Bill Cronon's book *Nature's Metropolis* was transformative for me, as it taught me the interconnectedness of rural and urban ar-eas, with significant implications for water, food, cities, wealth, and transportation. Bruce Hunt taught me a lot as a co-advisor to my undergraduate thesis in the mid-1990s and as author of *Pursuing Power and Light*, which was a useful reference for me on the history of the power sector. Larayne Dallas and Den-nis Trombatore at the University of Texas libraries helped me chase down some hard-to-find facts, for which I am grateful.

I have spent many hours over the last decade talking about energy with Roger Duncan, and his perspectives, especially about the future, have been very influential for me. I also fre-quently discuss energy with Russell Gold, who always has a unique insight to share. I want to thank my mother for intro-ducing me to Arthur C. Clarke's writings. His funny, optimistic views on the future, combined with his exasperation with the lack of creativity by his contemporaries, informed the Prologue and Epilogue.

The water chapter was based on a whole host of work from various students and collaborators, in particular Profes-sor Ashlynn Stillwell at the University of Illinois at Urbana-Champaign, Professor Kelly Sanders at the University of Southern California, Dr. Carey King at the University of Texas at Austin, and Professor Emily Grubert at Georgia Tech. I also collaborated with Professor Erica Belmont at the University of Wyoming, and Drs. F. Todd Davidson, Yael Glazer, and Emily

Beagle on water production from hydrocarbons. The importance of women's freedom and access to resources was written with Sheril Kirshenbaum, who also shows up as a co-author in the epilogue for rethinking energy education. I learned a lot from Martin Doyle and lean heavily on his book *The Source: How Rivers Made America and America Remade Its Rivers.* I also find Peter Gleick and David Sedlak to be go-to resources for all things water. Juan Miro taught me that the Aztec word for city is "water mountain."

The food chapter built extensively on work by Dr. Amanda Cuellar on the energy implications of animal waste and food waste. Dr. Kelly Sanders taught me the differences in the carbon intensities of wheat versus beef. Catherine Birney, Dr. F. Todd Davidson, and Katy Franklin were collaborators for the assessment of "foodprints." Isabella Gee taught me about meal delivery kits. Recent work with Dr. Charlie Upshaw and Heather Rose on the food-energy-water nexus makes an appearance in the discussion about urban farms. Dr. Colin Beal taught me a lot about life-cycle analysis of the food system. He and his father, Dr. Bill Beal, Professor Emeritus at Virginia Tech, provided thoughtful comments on the Prologue and the chapter on food, for which I am grateful. Professor Frank Mitloehner at University of California at Davis was the first person who really opened my eyes to the potential of reducing the environmental impacts of the food system by tackling its waste. I am still learning from him today. Professor Vijaya Nagarajan at the University of San Francisco taught me a new perspective on food waste and morality and gave me permission to quote one of her essays at length; I am grateful for both.

The transportation chapter includes references to work on mobility services by Dr. F. Todd Davidson, Gordon Tsai, and Zhenhong Lin at the Oak Ridge National Laboratory. Some of the top-level numbers on electrified transportation were quantified by Dr. Joshua Rhodes. I also learned from Dr. Kara Kockelman, who remains a world leader on modeling complex transportation systems. Impromptu conversations with Alan Lloyd and Dave Tuttle opened my eyes to some features of hydrogen or electrified transportation, for which I am thankful.

For the wealth chapter, my visit to Versailles was eye-opening. Though I had been there twice before, going there with Sandrine Voillet (http://sandrinevoillet.com) was a whole new experience, and I am very thankful for the rich details she taught me. I want to thank Professor Brian O'Gallachoir, whose social media postings tipped me off to the Electricity Supply Board archives in Ireland.

The cities chapter built on my partnership with Austin Energy and Pecan Street, where I learned a lot about the smart grid. I learned from Raj Bhattarai's holistic view of the world and tried to capture some of that perspective when discussing the problem of managing waste.

The security chapter includes extensive references to an article authored with Fred Beach about the use of oil and water to win World War II. I would like to especially thank Lt. Gen. (ret.) Ken Eickmann, who helped me track down some of the statistics on energy use by the DOD. The work of Professor Corey James at the US Military Academy at West Point also guided some of my thinking in a few key places.

The Epilogue built on what I have learned from Roger Duncan and incorporated some tidbits from Dr. Joshua Rhodes. Some of the work on rethinking energy education was written in collaboration with Sheril Kirshenbaum and is repurposed here. I would like to thank my daughter Evelyn for inspiring me with her acute observations.

In addition to this book, a companion documentary series was developed. The brainstorming and storyboarding sessions with the production team helped me shape the organization of the book. Marianne Shivers Gonzalez helped us make progress in the first few years with important inputs from Joyce Herring. Mat and Beth Hames of Alpheus Media and Juan Garcia of DISCO Learning Media were there from start to finish. I'd like to give a hat tip to Juan Garcia for suggesting the title *Power Trip: The Story of Energy*. They are all great to work with, and I am thankful for their partnership. Sponsors of that series helped motivate me to write the book, and include the Hewlett Foundation, the BQuest Foundation, Bill and Judy Bollinger, the Tiller Family Foundation, Mitch Julis, and others.

While the many collaborators helped me gain the knowledge I needed for this book, I would especially like to thank my patient wife Julia for putting up with me and nudging me along and for my kids for understanding that writing is a long process. I could not have done this without their support. They help give my life purpose.

# Notes

## PROLOGUE: THE STORY OF ENERGY

1. John N. Miksic and Geok Yian Goh, *Ancient Southeast Asia*, Routledge World Archaeology series (Florence, KY: Taylor and Francis, 2017), p. 14; and personal communication from John N. Miksic, October 30, 2018.

2. "Conservation of Energy," in Michael A. Gottlieb and Rudolf Pfeiffer, eds., *The Feynman Lectures on Physics,* vol. I, New Millennium Edition, California Institute of Technology, 2016 (1963), http://www .feynmanlectures.caltech.edu/I_04.html.

3. Arthur C. Clarke, *Profiles of the Future: An Inquiry into the Limits of the Possible* (New York: Popular Library, 1973).

4. Sun Ling Wang et al., "US Agricultural Productivity Growth: The Past, Challenges, and the Future," *Amber Waves* online magazine, US Department of Agriculture Economic Research Service, September 8, 2015, https://www.ers.usda.gov/amber-waves/2015/september/us-agricultural -productivity-growth-the-past-challenges-and-the-future/, accessed November 12, 2018.

5. Richard E. Smalley, "Future Global Energy Prosperity: The Terawatt Challenge," *MRS Bulletin* 30 (June 2005).

6. International Energy Agency, *Energy Access Outlook 2017: From Poverty to Prosperity*, World Energy Outlook Special Report (Paris, 2017), p. 3.

## CHAPTER 1: WATER

1. International Energy Agency, "Water Energy Nexus," excerpt from *World Energy Outlook 2016* (Paris, 2016).

2. An expanded discussion of water and civilization is covered in Michael E. Webber, *Thirst for Power: Energy, Water and Human Survival* (New Haven, CT: Yale University Press, 2016).

3. Fred Pearce, *When the Rivers Run Dry: Water—The Defining Crisis of the Twenty-First Century* (Boston: Beacon Press, 2006).

4. "Awash in Waste: Tradable Usage Rights Are a Good Tool for Tackling the World's Water Problems," *The Economist*, April 8, 2009.

5. A. T. Hodge, *Roman Aqueducts and Water Supply*, 2nd ed. (London: Bristol Classical Press, 2002).

6. Peter H. Gleick, "Water, Drought, Climate Change, and Conflict in Syria," *Weather, Climate, and Society* 6, no. 3 (July 2014).

7. Molly A. Walton, "Commentary: Energy Has a Role to Play in Achieving Universal Access to Clean Water and Sanitation," International Energy Agency, March 22, 2018, http://www.iea.org/newsroom /news/2018/march/commentary-energy-has-a-role-to-play-in-achieving -universal-access-to-clean-wate.html.

8. Martin Doyle, *The Source: How Rivers Made America and America Remade Its Rivers* (New York: W. W. Norton, 2018), p. 225.

9. Ibid.

10. Ibid., p. 222.

11. Ibid., p. 221.

12. Ibid., p. 232.

13. "1920s," Electricity Supply Board archives, https://esbarchives.ie /2016/03/10/esb-advertising-1920s/ Dublin, Ireland.

14. Doyle, *The Source*, p. 82.

15. B. Cameron Reed, "Kilowatts to Kilotons: Wartime Electricity Use at Oak Ridge," *History of Physics Newsletter* (American Physical Society) 12, no. 6 (Spring 2015), https://www.aps.org/units/fhp/newsletters /spring2015/upload/spring15.pdf, accessed May 27, 2018.

16. "League of Women Voters Through the Decades!" League of Women Voters, February 16, 2012, https://www.lwv.org/league-women -voters-through-decades.

17. Doyle, *The Source*, p. 197.

18. *Muscle Shoals*, directed by Greg "Freddy" Camalier, documentary, 2013.

19. Bonneville Power Administration, BPA Film Collection, Vol. One, 1939–1954, https://www.bpa.gov/news/AboutUs/History/Pages/Film-Collection-Vol-One.aspx, accessed July 5, 2018.

20. Doyle, *The Source*, p. 87.

21. Sardar Sarovar Narmada Nigam Ltd. (wholly owned government of Gujarat undertaking), http://sardarsarovardam.org, accessed August 4, 2018.

22. Xtankun Yang and X. X. Lu, "Ten Years of the Three Gorges Dam: A Call for Policy Overhaul," *Environmental Research Letters* 8 (2013).

23. Cutler Cleveland, "China's Monster Three Gorges Dam Is About to Slow the Rotation of the Earth," *Business Insider*, June 18, 2010, http://www.businessinsider.com/chinas-three-gorges-dam-really-will-slow-the-earths-rotation-2010-6, accessed July 5, 2018.

24. Yang and Lu, "Ten Years of the Three Gorges Dam."

25. Madhusree Mukerjee, "The Impending Dam Disaster in the Himalayas," *Scientific American*, July 14, 2015.

26. Jill Castellano, Tracy Loew, and Rosalie Murphy, "Cracks in the System: Oroville Crisis Highlights Risky Dams, Spotty Inspections Around U.S.," *Desert Sun*, USA Today Network, February 15, 2017.

27. International Rivers, "Grand Inga Dam, DR Congo," https://www.internationalrivers.org/campaigns/grand-inga-dam-dr-congo, accessed July 5, 2018.

28. Heather Rose, Charles Upshaw, and Michael E. Webber, "Evaluating Energy and Cost Requirements for Different Configurations of Off-Grid Rainwater Harvesting Systems," *Water* 10 (2018): 1024; DOI:10.3390/w10081024.

29. Doyle, *The Source*, p. 122.

30. Emily Grubert and Kelly T. Sanders, "Water Use in the United States Energy System: A National Assessment and Unit Process Inventory of Water Consumption and Withdrawals," *Environmental Science and Technology* 52, no. 11 (2018): 6695–6703, DOI: 10.1021/acs.est.8b00139.

31. E. L. Belmont et al., "Accounting for Water Formation from Hydrocarbon Fuel Combustion in Life Cycle Analyses," *Environmental Research Letters* 12 (2017), https://doi.org/10.1088/1748-9326/aa8390.

32. Jamie Satterfield, "180 New Cases of Dead or Dying Coal Ash Spill Workers, Lawsuit Says," *Knox News*, USA Today Network, March 28, 2018, https://www.knoxnews.com/story/news/crime/2018/03/28/tva-coal-ash

-spill-cleanup-roane-county-lawsuits-dead-dying-workers/458342002/, accessed March 30, 2018.

33. "Sewage Pollution of Water Supplies," *Engineering Record* 48 (August 1, 1903): 117, via Doyle, *The Source*, p. 176.

34. Ibid., p. 200.

35. United States Geological Survey, "The Water Cycle," http:// ga.water.usgs.gov/edu/watercycle.html, accessed December 31, 2014.

36. Peter H. Gleick, ed., *Water in Crisis: A Guide to the World's Fresh Water Resources* (New York: Oxford University Press, 1993).

37. Vaclav Smil, *Energy: A Beginner's Guide* (London: Oneworld Publications, 2006).

38. The original source for the estimates is Igor Shiklomanov's chapter, "World Fresh Water Resources," in Peter H. Gleick, ed., *Water in Crisis: A Guide to the World's Fresh Water Resources* (New York: Oxford University Press, 1993).

39. International Energy Agency, *Energy Access Outlook 2017: From Poverty to Prosperity*, World Energy Outlook Special Report (Paris, 2017), p. 3.

40. S. R. Kirshenbaum and M. E. Webber, "Liberation Power: What Do Women Need? Better Energy," *Slate*, November 4, 2013.

41. For a discussion of appliances and women, see Institute of Electrical and Electronics Engineers (IEEE) Global History Network, "Household Appliances and Women's Work," 2012; and "Fridges and Washing Machines Liberated Women, Study Suggests," *Science Daily*, March 13, 2009.

42. Elaine Tyler May, *Homeward Bound* (New York: Basic Books, 1988).

43. For a good history of water and wastewater infrastructure and treatment systems, see David Sedlak, *Water 4.0: The Past, Present, and Future of the World's Most Vital Resource* (New Haven, CT: Yale University Press, 2014).

44. Margaret B. Freeman, *The Unicorn Tapestries*, exhibition catalogue, the Cloisters, New York, 1976.

45. Doyle, *The Source*, p. 171.

46. Sedlak, *Water 4.0*.

47. For information on global access to water and wastewater, the United Nations is a reliable source. For information on global water stress, Vörösmarty's work is the standard-bearer: C. J. Vörösmarty et al., "Global Threats to Human Water Security and River Biodiversity," *Nature* 467, September 30, 2010. In addition, the Pacific Institute produces

a series of biennial reports on freshwater resources with convenient summaries of water data and in-depth analyses on water topics, including availability, access, policies, and technologies: Peter Gleick et al., *The World's Water*, vol. 8 (Washington, DC: Island Press, 2014; and prior volumes in prior years).

48. Jeff Goodell, *Big Coal: The Dirty Secret Behind America's Energy Future* (New York: Houghton Mifflin Harcourt, 2006).

49. Peter H. Gleick, *Bottled and Sold: The Story Behind Our Obsession with Bottled Water* (Washington, DC: Island Press, 2010).

50. Peter H. Gleick and Heather S. Cooley, "Energy Implications of Bottled Water," *Environmental Research Letters* 4 (2009).

51. Lizzie Dearden, "Venezuela Energy Crisis: President Tells Women to Stop Using Hairdryers and Go with 'Natural' Style to Save Electricity," *The Independent*, April 9, 2016.

52. Mark Raby, "Tea Time in Britain Causes Predictable, Massive Surge in Electricity Demand," *Geek-Cetera*, January 7, 2013, https://www.geek .com/geek-cetera/tea-time-in-britain-causes-predictable-massive-surge -in-electricity-demand-1535023/, accessed November 13, 2018.

53. Karl Smallwood, "Does The UK Really Experience Massive Power Surges When Soap Operas Finish from People Making Tea?," Today I Found Out, May 28, 2017, http://www.todayifoundout.com/index.php /2017/05/uk-really-experience-power-surges-soap-operas-finish/, accessed November 13, 2018.

54. C. Song et al., "An Analysis on the Energy Consumption of Circulating Pumps of Residential Swimming Pools for Peak Load Management," *Applied Energy* 195 (2017): 1–12, https://www.sciencedirect.com /science/article/pii/S0306261917302416.

55. International Energy Agency, "Water Energy Nexus," excerpt from *World Energy Outlook 2016* (Paris, 2016).

56. K. T. Sanders and M. E. Webber, "Evaluating the Energy Consumed for Water Use in the United States," *Environmental Research Letters* 7 (2012).

57. Norman Chan, "What SpaceX's Dragon Brought to the International Space Station," *Tested*, May 25, 2012, http://www.tested.com /science/space/44509-what-spacexs-dragon-brought-to-the-international -space-station/, accessed January 3, 2015.

58. M. E. Webber, D. S. Baer, and R. K. Hanson, "Ammonia Monitoring Near 1.5 μm with Diode Laser Absorption Sensors," *Applied Optics*

40, no. 12 (2001); M. E. Webber et al., "Measurements of $NH_3$ and $CO_2$ with Distributed-Feedback Diode Lasers Near 2 μm in Bioreactor Vent Gases," *Applied Optics* 40, no. 24 (2001).

59. C. M. Chini et al., "Quantifying Energy and Water Savings in the US Residential Sector," *Environmental Science and Technology* 50, no. 17 (2016): 9003–9012, DOI: 10.1021/acs.est.6b01559.

## CHAPTER 2: FOOD

1. C. I. Birney et al., "An Assessment of Individual Foodprints Attributed to Diets and Food Waste in the United States," *Environmental Research Letters* 12, no. 10 (2017).

2. A. D. Cuellar and M. E. Webber, "Wasted Food, Wasted Energy: The Embedded Energy in Food Waste in the United States," *Environmental Science and Technology* 44, no. 16, July 21, 2010.

3. Brian Fagan, *Elixir: A History of Water and Humankind* (London: Bloomsbury Press, 2011).

4. Brian Fagan, *The Great Warming: Climate Change and the Rise and Fall of Civilizations* (London: Bloomsbury Press, 2009).

5. National Academies of Sciences, Engineering, and Medicine, *Nutrient Requirements of Beef Cattle*, 8th rev. ed. (Washington, DC: National Academies Press, 2016), https://doi.org/10.17226/19014.

6. Martin Doyle, *The Source: How Rivers Made America and America Remade Its Rivers* (New York: W. W. Norton, 2018).

7. US Department of Agriculture, *Equine 2005, Part II: Changes in the US Equine Industry, 1998–2005*, March 2007. USDA-APHIS-VS, CEAH, Fort Collins, CO, N452-0307.

8. Jay F. Martin et al., "Energy Evaluation of the Performance and Sustainability of Three Agricultural Systems with Different Scales and Management," *Agriculture, Ecosystems, and Environment* 115 (2006): 128–140; Dazhong Wen and D. Pimentel, "Energy Inputs in Agricultural Systems in China," *Agriculture, Ecosystems, and Environment* 11 (1984): 29–35.

9. Minnesota Historical Society, Mill City Museum, http://www.mnhs.org/millcity/learn, accessed October 27, 2018.

10. D. Pimentel, *Food, Energy, and Society* (Hoboken, NJ: Taylor and Francis, 2007).

11. Horizon Farms, Nutritional Label, https://www.horizon.com/products/milk/whole-milk, accessed July 7, 2018.

12. As per Sandrine Voillet, there are several books referring to the *glacières*, or ice pits, such as Nicolas Jacquet, *Secrets of Versailles* (Paris: Parigramme, 2011), 177. This French Wikipedia article also has some information: https://fr.wikipedia.org/wiki/Glaci%C3%A8res_de_Versailles.

13. Emily Badger, "What Happens When the Richest US Cities Turn to the World?" *New York Times*, December 22, 2017, https://www.nytimes.com/2017/12/22/upshot/the-great-disconnect-megacities-go-global-but-lose-local-links.html?_r=0.

14. William Cronon, *Nature's Metropolis: Chicago and the Great West* (New York: W. W. Norton, 1991), p. 225.

15. Ibid., p. 231.

16. Mark Kurlansky, *Salt: A World History* (New York: Walker and Company, 2002).

17. "*Industries of the British Empire* (1933)," The Art of Rockefeller Center, Rockefeller Center, accessed December 7, 2018, https://www.rockefellercenter.com/art-and-history/art/industries-of-the-british-empire/.

18. James L. Sweeney, *Energy Efficiency: Building a Clean, Secure Economy*, Hoover Institution Press Publication no. 668 (Stanford, CA: Hoover Institution at Leland Stanford Junior University, 2016).

19. US Department of Energy, Energy Information Administration, "Residential Energy Consumption Survey (RECS) 2015," released 2017, https://www.eia.gov/consumption/residential/index.php, accessed July 8, 2018.

20. Amanda Cuellar and Michael E. Webber, "Wasted Food, Wasted Energy: The Embedded Energy in Food Waste in the United States," *Environmental Science and Technology* 44 (2010): 6464–6469.

21. Winifred Gallagher, *House Thinking: A Room-by-Room Look at How We Live* (New York: HarperCollins, 2006).

22. C. M. Saunders, A. Barber, and Gregory J. Taylor, "Food Miles—Comparative Energy/Emissions Performance of New Zealand's Agriculture Industry," 2006, AERU Research Report series 344, Agribusiness and Economics Research Unit (AERU), Lincoln University, New Zealand, https://researcharchive.lincoln.ac.nz/handle/10182/125, accessed November 13, 2018.

23. Tyler Colman and Pablo Päster, "Red, White, and 'Green': The Cost of Greenhouse Gas Emissions in the Global Wine Trade," *Journal of Wine Research* 20, no. 1 (2009): 15–26, DOI: 10.1080/09571260902978493.

24. Intergovernmental Panel on Climate Change, "2014: Summary for Policymakers," in *Climate Change 2014: Mitigation of Climate Change. Contribution of Working Group III to the Fifth Assessment Report of the Intergovernmental Panel on Climate Change*, O. Edenhofer et al., eds. (Cambridge, UK: Cambridge University Press, 2014).

25. Lighting is responsible for about 7 percent of total electricity consumption or about 3 percent of total energy consumption. US Energy Information Administration, FAQ, "How Much Electricity Is Used for Lighting in the United States?," February 9, 2018, https://www.eia.gov /tools/faqs/faq.php?id=99&t=3, accessed July 8, 2018.

26. Christopher L. Weber and H. Scott Matthews, "Food-Miles and the Relative Climate Impacts of Food Choices in the United States," *Environmental Science and Technology* 42, no. 10 (2008): 3508–3513, DOI: 10.1021/es702969f.

27. K. T. Sanders and M. E. Webber, "A Comparative Analysis of the Greenhouse Gas Emissions Intensity of Wheat and Beef in the United States," *Environmental Research Letters* 9 (2014).

28. Weber and Matthews, "Food-Miles."

29. Michael Pollan, "Farmer-in-Chief," *New York Times Magazine*, October 12, 2008.

30. David S. Ludwig et al., "Dietary Carbohydrates: Role of Quality and Quantity in Chronic Disease," *BMJ* 361 (June 13, 2018): k2340, https://doi.org/10.1136/bmj.k2340.

31. David J. Suskind et al., "Clinical and Fecal Microbial Changes with Diet Therapy in Active Inflammatory Bowel Disease," *Journal of Clinical Gastroenterology* 52, no. 2 (February 2018): 155–163, DOI: 10.1097 /MCG.0000000000000772.

32. Azeen Ghorayshi, "Too Big to Chug: How Our Sodas Got So Huge," *Mother Jones*, June 25, 2012, http://www.motherjones.com/media /2012/06/supersize-biggest-sodas-mcdonalds-big-gulp-chart/.

33. E. Barclay, J. Belluz, and J. Zarracina, "It's Easy to Become Obese in America. These 7 Charts Explain Why," October 13, 2017, https://www .vox.com/2016/8/31/12368246/charts-explain-obesity.

34. Steve Connor, "Super Size Me: How the Last Supper Became a Banquet over 1,000 Years," *The Independent*, March 24, 2010, http:// www.independent.co.uk/news/science/super-size-me-how-the-last -supper-became-a-banquet-over-1000-years-1926159.html.

35. Michael Kahn, "Obesity Contributes to Global Warming: Study," Reuters, May 15, 2008.

36. Food and Agriculture Organization (FAO) of the United Nations, "Food Loss and Food Waste," http://www.fao.org/food-loss-and-food-waste/en/, accessed July 8, 2018.

37. A. D. Cuellar and M. E. Webber, "Wasted Food, Wasted Energy: The Embedded Energy in Food Waste in the United States," *Environmental Science and Technology*, 44, no. 16, July 21, 2010.

38. Here is an example of an ugly fruit and vegetable campaign: http://www.endfoodwaste.org/ugly-fruit---veg.html.

39. Vijaya Nagarajan, *Feeding a Thousand Souls: Women, Ritual and Ecology in India—An Exploration of the Kolam* (New York: Oxford University Press, 2018), pp. 240–242.

40. Kurlansky, *Salt: A World History*.

41. Adam Thomson, "'Tortilla Riots' Give Foretaste of Food Challenge," *Financial Times*, October 12, 2010.

42. Ines Perez, "Climate Change and Rising Food Prices Heightened Arab Spring," ClimateWire via *Scientific American*, March 4, 2013.

43. "Food and the Arab Spring: Let Them Eat Baklava," *The Economist*, May 17, 2012.

44. The Smithsonian and USDA created exhibits to show a collection of these posters. See, for example, Cory Bernat, "An Exhibition of Posters," http://www.good-potato.com/beans_are_bullets/index.html, 2010; and Amanda Fiegl, *Smithsonian Magazine*, May 28, 2010, https://www.smithsonianmag.com/arts-culture/american-food-posters-from-world-war-i-and-ii-89453240/.

45. USDA, Economic Research Service, "US Domestic Corn Use," chart in "Corn and Other Feedgrains," May 15, 2018, https://www.ers.usda.gov/topics/crops/corn-and-other-feedgrains/background/, accessed November 13, 2018.

46. Joel K. Bourne Jr., "Green Dreams," *National Geographic*, October 2007.

47. Terry Macalister, "Biofuel Bonanza Not So Sweet for Brazil's Sugar Cane Cutters," *The Guardian*, June 4, 2008, https://www.theguardian.com/environment/2008/jun/04/biofuels.oil, accessed November 13, 2018.

48. "Coffee-Enhanced Fuel Set to Power London Buses," Agence France-Presse, November 17, 2017, https://www.pri.org/stories/2017-11-17/coffee-enhanced-fuel-set-power-london-buses, accessed April 30, 2018.

49. Austin Water, 'Dillo Dirt website, http://www.austintexas.gov/dillodirt, accessed July 8, 2018.

50. Pollan, "Farmer-in-Chief."

51. M. E. Webber et al., "Agricultural Ammonia Sensor Using Diode Lasers and Photoacoustic Spectroscopy," *Measurement Science and Technology* 16 (2005): 1547–1553.

52. Amplifier-Enhanced Optical Analysis System and Method, Patent No. 7,064,329 (2006), described in Webber et al., "Agricultural Ammonia Sensor."

53. Amanda D. Cuellar and Michael E. Webber, "Cow Power: The Energy and Emissions Benefits of Converting Manure to Biogas," *Environmental Research Letters* 3 (July 2008).

54. Bathina Chaitanya et al., "Biomass-Gasification-Based Atmospheric Water Harvesting in India," *Energy* 165 (2018): 610–621.

55. Colin M. Beal et al., "Energy Return on Investment for Algal Biofuel Production Coupled with Wastewater Treatment," *Water Environment Research* 84, no. 9 (2012).

56. Judith Lewis Mernit, "How Eating Seaweed Can Help Cows to Belch Less Methane," YaleEnvironment360, July 2, 2018, https://e360 .yale.edu/features/how-eating-seaweed-can-help-cows-to-belch-less -methane, accessed July 8, 2018.

57. Paul Horn, "Infographic: Why Farmers Are Ideally Positioned to Fight Climate Change," Inside Climate News, October 24, 2018, https://insideclimatenews.org/news/24092018/infographic-farm-soil -carbon-cycle-climate-change-solution-agriculture, accessed October 28, 2018.

58. Caitlin Dewey, "You're About to See a Big Change to the Sell-By Dates on Food," *Washington Post*, February 16, 2017.

59. "The Dating Game: How Confusing Labels Land Billions of Pounds of Food in the Trash," Natural Resources Defense Council Issue Brief, September 2013, Issue Brief 13-9-A, https://www.nrdc.org/sites/default /files/dating-game-IB.pdf, accessed November 13, 2018.

60. ReFED, "A Roadmap to Reduce US Food Waste by 20 Percent," report, 2016, https://www.refed.com/downloads/ReFED_Report_2016.pdf, accessed November 13, 2018.

## CHAPTER 3: TRANSPORTATION

1. Arthur C. Clarke, *Profiles of the Future: An Inquiry into the Limits of the Possible* (New York: Popular Library, 1973).

2. Martin Doyle, *The Source: How Rivers Made America and America Remade Its Rivers* (New York: W. W. Norton, 2018), p. 31.

3. Doris Kearns Goodwin, *Team of Rivals* (New York: Simon & Schuster, 2006).

4. William Cronon, *Nature's Metropolis: Chicago and the Great West* (New York: W. W. Norton, 1991).

5. Ibid.

6. Michael V. Hazel, "Browder's Springs," Texas State Historical Association, https://tshaonline.org/handbook/online/articles/rpb04, accessed April 21, 2018; "A History of Railroads in Dallas," Museum of the American Railroad, http://www.museumoftheamericanrailroad.org/learn/ahistoryofrailroadsinnorthtexas.aspx, accessed April 21, 2018.

7. Stanley Steamer, Stanley Motor Carriages Technical Information, http://www.stanleymotorcarriage.com/GeneralTechnical/General Technical.htm, accessed April 30, 2018.

8. See for example Kate VanDyke, *Fundamentals of Petroleum*, 4th ed. (Austin, TX: Petroleum Extension Service, 1997), p. 211.

9. Douglas A. McIntyre, "America's Biggest Companies, Then and Now (1955 to 2010)," 24/7 Wall St. blog, September 21, 2010. https://247wallst.com/investing/2010/09/21/americas-biggest-companies-then-and-now-1955-to-2010/, accessed November 13, 2018.

10. David Halberstam, *The Fifties* (New York: Ballantine Books, 1993).

11. Stephen A. Ambrose, "Eisenhower: Soldier and President" (New York: Simon & Schuster, 1991).

12. The History Channel, "The Interstate Highway System," A&E Television Networks, https://www.history.com/topics/interstate-highway-system, accessed November 13, 2018.

13. David Halberstam, *The Fifties* (New York: Ballantine Books, 1993).

14. Philip S. Schmidt and John R. Howell, "Historical Background on the Brayton Cycle and Development of Gas Turbine Engines," American Society of Engineering Education, Centennial Conference, June 20–23, 1993.

15. Quoted in John Golley, *Whittle: The True Story* (Washington, DC: Smithsonian Institution Press, 1987).

16. Clarke, *Profiles of the Future*.

17. Schmidt and Howell, "Historical Background on the Brayton Cycle and Development of Gas Turbine Engines."

18. Kenneth Chang, "25 Years Ago, NASA Envisioned Its Own 'Orient Express,'" *New York Times*, October 20, 2014.

19. Bruce Hunt, *Pursuing Power and Light* (Baltimore: Johns Hopkins University Press, 2010).

20. Joshua D. Rhodes and Michael E. Webber, "The Solution to America's Energy Waste Problem," *Fortune*, December 18, 2017, http://fortune .com/2017/12/18/electrification-energy-u-s-economy/, accessed April 30, 2018.

21. Jonathan Mahler, "The Case for the Subway," *New York Times Magazine*, January 3, 2018.

22. Robert M. Salter, *The Very High Speed Transit System*, RAND report P-4874, Santa Monica, CA, August 1972.

23. Emilia Simeonova et al., "Congestion Pricing, Air Pollution, and Children's Health," National Bureau of Economic Research, Working Paper 24410, March 2018.

24. Justin Gillis and Hal Harvey, "Cars Are Ruining Our Cities," *New York Times*, April 25, 2018.

25. Eric Jaffe, "California's DOT Admits That More Roads Mean More Traffic," CityLab, November 11, 2015, https://www.citylab.com /transportation/2015/11/californias-dot-admits-that-more-roads-mean -more-traffic/415245/, accessed May 1, 2018.

26. Simeonova, "Congestion Pricing, Air Pollution, and Children's Health."

27. Ibid.

28. Much of this section was originally published in Michael E. Webber, "A New Age of Rail," *Mechanical Engineering*, February 2018.

29. Bureau of Transportation Statistics, *Freight Facts and Figures 2015*, (US Department of Transportation, 2015), https://www.bts.gov/sites/bts .dot.gov/files/legacy/FFF_complete.pdf.

30. Brian A. Weatherford, Henry H. Willis, and David S. Ortiz, *The State of US Railroads: A Review of Capacity and Performance Data* (Santa Monica, CA: RAND Corporation, 2008), p. 14.

31. Bureau of Transportation Statistics, *Freight Facts and Figures 2015*, p. 23.

32. Ibid., p. 65.

33. Weatherford, Willis, and Ortiz, *The State of US Railroads*, p. 12.

34. Ibid., 38.

35. For $CO_2$ emissions, see Bureau of Transportation Statistics, *Freight Facts and Figures 2015*, p. 88; for percentage of ton-miles, Department of Transportation, Federal Railroad Administration, *National Rail Plan Progress Report*, September 2010, p. 14.

36. Department of Transportation, Federal Railroad Administration, *National Rail Plan Progress Report*, 2010.

37. Bureau of Transportation Statistics, *Freight Facts and Figures 2015*.

38. Ibid., p. 4.

39. Ibid., pp. 75 and 77.

40. Michael Anderson and Maximilian Auffhammer, "Pounds That Kill: The External Costs of Vehicle Weight," *Review of Economic Studies* 81, no. 2 (April 1, 2014): 535–571, https://doi.org/10.1093/restud/rdt035.

41. Bureau of Transportation Statistics, *Freight Facts and Figures 2015*, p. 75.

42. K. A. Small, C. Winston, and C. A. Evans, *Road Work: A New Highway Pricing and Investment Policy* (Washington, DC: Brookings Institution, 1989).

43. M. E. Webber, "Coal Country Isn't Coming Back," *New York Times*, November 12, 2016.

44. Alexander E. MacDonald et al., "Future Cost-Competitive Electricity Systems and Their Impact on US $CO_2$ Emissions," *Nature Climate Change* 6 (2016): 526–531.

45. Bureau of Transportation Statistics, *Freight Facts and Figures 2015*, p. 72.

46. M. E. Webber, "How to Overhaul the Gas Tax," *New York Times*, December 23, 2013.

47. International Energy Agency, *Railway Handbook 2017: Energy Consumption and CO2 Emissions* (Paris, 2017).

48. "Artificial Intelligence and Life in 2030," One Hundred Year Study on Artificial Intelligence, panel discussion, Stanford University, September 2016, https://ai100.stanford.edu/sites/default/files/ai_100_report_0831fnl.pdf, accessed November 13, 2018.

49. Daniel M. Kammen and Deborah A. Sunter, "City-Integrated Renewable Energy for Urban Sustainability," *Science* 352, no. 6288, May 20, 2016.

50. Christopher Monsere et al., "Lessons from the Green Lanes: Evaluating Protected Bike Lanes in the US," National Institute for Transportation and Communities (NITC), Portland, Oregon, 2014, http://trec.pdx.edu/research/project/583/Lessons_from_the_Green_Lanes:_Evaluating_Protected_Bike_Lanes_in_the_U.S.

51. F. Todd Davidson et al., "An Analytical Framework for Quantifying the Economic Value of Mobility Services," *in review*. We built an online interactive calculator that lets people figure out whether it's cheaper to own a car or use a transportation network company service

based on their mobility requirements, the value of their time, and so on: www.rideordrive.org.

52. Donald Shoup, "Free Parking or Free Markets," *ACCESS: The Magazine of UCTC* (Spring 2011), https://www.researchgate.net /publication/285889431, accessed November 13, 2018.

53. Patrick Sawer, "Motorists Spend Four Days a Year Looking for a Parking Space," *The Telegraph*, February 1, 2017, http://www.telegraph.co.uk /news/2017/02/01/motorists-spend-four-days-year-looking-parking-space/.

54. Daniel J. Fagnant and Kara M. Kockelman, "The Travel and Environmental Implications of Shared Autonomous Vehicles, Using Agent-Based Model Scenarios," *Transportation Research Part C: Emerging Technologies* 40 (2014): 1–13, http://www.sciencedirect.com/science /article/pii/S0968090X13002581.

55. Kendra L. Smith, "How to Ensure Smart Cities Benefit Everyone," *The Conversation*, October 31, 2016.

56. John Morell, "Train Versus Traffic," *Mechanical Engineering*, February 2017.

57. J. D. Power, "New Vehicle Retail Sales Pace to Decline for Fourth Consecutive Month," July 27, 2017, http://www.jdpower.com/press -releases/jd-power-and-lmc-automotive-forecast-july-2017; US Department of Transportation, Bureau of Transportation Statistics, "National Household Travel Survey Daily Travel Quick Facts," https://www.bts .gov/statistical-products/surveys/national-household-travel-survey -daily-travel-quick-facts, last updated May 31, 2017.

58. This section's concepts were originally published in F. Todd Davidson and Michael E. Webber, "To Uber or Not? Why Car Ownership May No Longer Be a Good Deal," *The Conversation*, October 15, 2017.

59. "National Household Travel Survey Daily Travel Quick Facts," Bureau of Transportation Statistics, US Department of Transportation, updated May 31, 2017, https://www.bts.gov/statistical-products/surveys /national-household-travel-survey-daily-travel-quick-facts; "Average Annual Miles per Driver by Age Group," Office of Highway Policy Information, Federal Highway Administration, US Department of Transportation, last modified March 29, 2018, https://www.fhwa.dot.gov/ohim/onh00 /bar8.htm.

60. Matt McFarland, "How Free Self-Driving Car Rides Could Change Everything," CNN, September 1, 2017, http://money.cnn.com /2017/09/01/technology/future/free-transportation-self-driving-cars /index.html, accessed May 2, 2018.

61. Gillian Tett, "US Truck Driver Shortage Points to Bigger Problems," *Financial Times*, April 8, 2018.

## CHAPTER 4: WEALTH

1. Rose George, *The Big Necessity: The Unmentionable World of Human Waste and Why It Matters* (New York: Picador, 2008).

2. Chelsea Follett, "Technological Progress Freed Children from Hard Labor," CATO at Liberty, July 2, 2018, https://www.cato.org/blog /technological-progress-freed-kids-hard-labor, accessed August 19, 2018.

3. Wolfgang Schivelbusch, *Disenchanted Night: The Industrialization of Light in the Nineteenth Century* (Berkeley: University of California Press, 1995).

4. Martin Doyle, *The Source: How Rivers Made America and America Remade Its Rivers* (New York: W. W. Norton, 2018).

5. Bryan Burrough, *The Big Rich: The Rise and Fall of the Greatest Texas Oil Fortunes* (New York: Penguin, 2009).

6. Arman Shehabi et al., *United States Data Center Energy Usage Report*, Lawrence Berkeley National Laboratory, US Department of Energy, June 2016, LBNL-1005775.

7. Anmar Frangoul, "9.8 Million People Employed by Renewable Energy, According to New Report," CNBC, May 24, 2017, https://www .cnbc.com/2017/05/24/9-point-8-million-people-employed-by-renew able-energy-according-to-new-report.html; Martin Tiller, "The Fossil Fuel Industry May Not Help the Planet, but It Employs Millions," July 9, 2014, "Oil and Gas Employment in the United States," https://oilprice .com/Energy/Energy-General/The-Fossil-Fuel-Industry-May-Not-Help -the-Planet-But-It-Employs-Millions.html.

8. "Did You Know? Key Facts About the Shannon Scheme," Electricity Supply Board archives, https://esbarchives.ie/test-ss-key-facts-page/ Dublin, Ireland.

9. "1920s," Electricity Supply Board archives, https://esbarchives .ie/2016/03/10/esb-advertising-1920s/ Dublin, Ireland.

10. Jeff Goodell, *Big Coal: The Dirty Secret Behind America's Energy Future* (New York: Houghton Mifflin Harcourt, 2006).

11. Georges Vigarello, *Le propre et le sale: L'hygiène du corps depuis le Moyen Âge*, trans. Sandrine Voillet (Paris: Du Seuil, 1985), p. 116.

12. Lindsey Fitzharris, *The Butchering Art: Joseph Lister's Quest to Transform the Grisly World of Victorian Medicine* (New York: Scientific American, 2017).

13. Electricity Supply Board archives, "Prints for Adverts: 1930s," https://esbarchives.ie/2016/03/09/esb-advertising-1920s-and-1930s/, Dublin, Ireland.

14. Bonneville Power Administration, *Hydro*, BPA Film Collection, Vol. One, 1939–1954, https://www.bpa.gov/news/AboutUs/History/Pages /Film-Collection-Vol-One.aspx, accessed July 5, 2018.

## CHAPTER 5: CITIES

1. David Sedlak, *Water 4.0: The Past, Present, and Future of the World's Most Vital Resource* (New Haven: Yale University Press, 2015).

2. Ed Crooks, "Mayors Look to Tackle Climate Change at City Level," *Financial Times*, December 1, 2016.

3. Paris Agreement summary, European Commission, accessed December 10, 2018, https://ec.europa.eu/clima/policies/international /negotiations/paris_en; United Nations, "Cities Striving to Play Key Role in Implementing Paris Agreement," November 10, 2016, http://www.un .org/sustainabledevelopment/blog/2016/11/cities-striving-to-play-key -role-in-implementing-paris-agreement/.

4. Wolfgang Schivelbusch, *Disenchanted Night: The Industrialization of Light in the Nineteenth Century* (Berkeley: University of California Press, 1995), p. 7.

5. Etablissement Public du Musée et Domaine National de Versailles, "La galerie des Glaces: Le Brun, maître d'œuvre," October 2007, http:// www.chateauversailles.fr/resources/pdf/fr/public-spe/aide_visite_galeries glaces.pdf, accessed August 27, 2018.

6. "How the Cost of Light Fell by a Factor of 500,000," Human Progress, February 15, 2017, https://humanprogress.org/article.php?p=495, accessed August 27, 2018.

7. Daniel Yergin, "The First War to Run on Oil," *Wall Street Journal*, August 14, 2014.

8. Electricity Supply Board archives, "Print Adverts by Decade: 1930s," carousel view, https://esbarchives.ie/2016/03/09/esb-advertising-1920s -and-1930s/#jp-carousel-1303, Dublin, Ireland.

9. Rowan Moore, "An Inversion of Nature: How Air Conditioning Created the Modern City," *The Guardian*, August 14, 2018, https://www .theguardian.com/cities/2018/aug/14/how-air-conditioning-created -modern-city, accessed August 19, 2018.

10. Molly Ivins, "King of Cool," *Time*, December 7, 1998.

11. Moore, "An Inversion of Nature."

12. Wolfgang Schivelbusch, *Disenchanted Night: The Industrialization of Light in the Nineteenth Century* (Berkeley: University of California Press, 1995).

13. Schivelbusch, *Disenchanted Night*, p. 89.

14. Julia Hider, "The Lost American Museum That Had It All," Messy Nessy, January 12, 2017, http://www.messynessychic.com/2017/01/12/the-lost-american-museum-that-had-it-all/, accessed August 23, 2018.

15. Philip B. Kunhardt Jr., Philip B. Kunhardt III, and Peter W. Kunhardt, *P. T. Barnum: America's Greatest Showman* (New York: Knopf, 1995), p. 244.

16. Schivelbusch, *Disenchanted Night*, p. 32.

17. Laura Freeman, "The Most Magical Job in Britain: Enchanting Story of Our Last Gas Street Lights . . . " November 24, 2014, http://www.dailymail.co.uk/news/article-2848038/The-magical-job-Britain-Enchanting-story-gas-street-lights-five-men-burning-just-did-Dickens-day.html.

18. Nicholas St. Fleur, "Illuminating the Effects of Light Pollution," *New York Times*, April 7, 2016.

19. "Light Pollution Effects on Wildlife and Ecosystems," International Dark Sky Association, http://darksky.org/light-pollution/wildlife/, accessed August 12, 2018.

20. H. Poot, B. J. Ens, H. de Vries, M. A. H. Donners, M. R. Wernand, and J. M. Marquenie, "Green Light for Nocturnally Migrating Birds," *Ecology and Society* 13, no. 2 (2008): 47, http://www.ecologyandsociety.org/vol13/iss2/art47/; Ian Randall, "Blue Lights Could Prevent Bird Strikes," *Science*, April 13, 2015, https://www.sciencemag.org/news/2015/04/blue-lights-could-prevent-bird-strikes; US Fish and Wildlife Service Division of Migratory Bird Management, *Reducing Bird Collisions with Buildings and Building Glass: Best Practices* (Falls Church, VA: US Fish and Wildlife Service, 2016), https://www.fws.gov/migratorybirds/pdf/management/reducingbirdcollisionswithbuildings.pdf.

21. Austin Troy, *The Very Hungry City* (New Haven: Yale University Press, 2012).

22. William Cronon, *Nature's Metropolis: Chicago and the Great West* (New York: W. W. Norton, 1991).

23. Allen MacDuffie, "Dickens and Energy," Faculty Seminar on British Studies, Harry Ransom Humanities Research Center, Austin, TX,

November 15, 2013; MacDuffie, *Victorian Literature, Energy, and the Ecological Imagination* (Cambridge: Cambridge University Press, 2014).

24. "A Thaw in the Streets of London," *The Illustrated London News*, February 25, 1865, 184–185, http://www.victorianweb.org/science/health /streets.jpg.

25. Mathias Basner et al., "Auditory and Non-auditory Effects of Noise on Health," *Lancet* 383, no. 9925 (April 12, 2014): 1325–1332, https:// www.ncbi.nlm.nih.gov/pmc/articles/PMC3988259/, accessed August 12, 2018.

26. Richard Stone, "Counting the Cost of London's Killer Smog," *Science* 298, no. 5601 (December 13, 2002): 2106–2107, DOI: 10.1126 /science.298.5601.2106b.

27. Martin Doyle, *The Source: How Rivers Made America and America Remade Its Rivers* (New York: W. W. Norton, 2018).

28. Cronon, *Nature's Metropolis*.

29. Doyle, *The Source*, p. 165.

30. Cronon, *Nature's Metropolis*, p. 12 ("blanketing the prairie"); p. 142 ("In nature's economy").

31. Ibid., p. 341.

32. Emily Badger, "What Happens When the Richest US Cities Turn to the World?," *New York Times*, December 22, 2017, https://www .nytimes.com/2017/12/22/upshot/the-great-disconnect-megacities-go -global-but-lose-local-links.html?_r=0.

33. Much of this section on reducing waste was originally published in Michael E. Webber, "Tapping the Trash," *Scientific American*, July 2017.

34. Hong Yang et al., "The Crushing Weight of Urban Waste," *Science* 351, no. 6274 (February 12, 2016): 674.

35. Warren Cornwall, "Sewage Sludge Could Contain Millions of Dollars Worth of Gold," *Science*, January 16, 2015, http://www.sciencemag.org /news/2015/01/sewage-sludge-could-contain-millions-dollars-worth-gold.

36. "From Waste to Energy: Zürich," My Switzerland, http://www.my switzerland.com/en-us/zuerich-warmth.html; John B. Kitto Jr. and Larry A. Hiner, "Clean Power From Burning Trash," *Mechanical Engineering*, February 2017.

37. Alex C. Breckel, John R. Fyffe, and Michael E. Webber, "Trash-to-Treasure: Turning Non-Recycled Waste into Low-Carbon Fuel," *Earth* 57, no. 8: 42.

38. Kenneth R. Weiss, "Vancouver's Green Dream," *Science* 352, no. 6288 (May 20, 2016): 918–921.

39. Friotherm AG, "Energy from sewage water – District heating and district cooling in Sandvika, with 2 Unitop® 28C heat pump units," sales brochure, uploaded November 2017, https://www.friotherm.com /wp-content/uploads/2017/11/sandvika_e005_uk.pdf.

40. R. P. Siegel, "A Natural Fit," *Mechanical Engineering*, May 2016.

41. Ayona Datta, "Will India's Experiment with Smart Cities Tackle Poverty—Or Make It Worse?" The Conversation, January 27, 2016, https://theconversation.com/will-indias-experiment-with-smart-cities -tackle-poverty-or-make-it-worse-53678.

42. Ali M. Sadeghioon et al., "SmartPipes: Smart Wireless Sensor Networks for Leak Detection in Water Pipelines," *Journal of Sensor and Actuator Networks* 3 (2014).

43. Kendra L. Smith, "How to Ensure Smart Cities Benefit Everyone," *The Conversation*, October 31, 2016.

44. Ibid.

## CHAPTER 6: SECURITY

1. Peter Gleick, "Water and US National Security," War Room, United States Army War College, June 15, 2017; Mark Kurlansky, *Salt: A World History* (New York: Walker and Company, 2002).

2. David Marsh, "Oil Remains the Driving Force of the Persian Gulf War," *Washington Post*, January 23, 1991.

3. Transcript of President George H.W. Bush's Address on the End of the Gulf War, *New York Times*, March 7, 1991.

4. Basia Rosenbaum, "The Battle for Arctic Oil," *Harvard International Review*, March 9, 2015.

5. Daniel Yergin, "The First War to Run on Oil," *Wall Street Journal*, August 14, 2014.

6. Ibid.

7. Robert Paulus, "A Full Partner—Logistics and the Joint Force," *Army Logistician*, July 1, 2003.

8. Much of this section is based on an article I co-authored with Dr. Fred Beach at the University of Texas at Austin: Fred C. Beach and Michael E. Webber, "How Oil and Water Helped the US Win World War II," *Earth* 56, no. 3: 34.

9. "US Ship Force Levels," Naval History and Heritage Command, published Friday, November 17, 2017, https://www.history.navy.mil /research/histories/ship-histories/us-ship-force-levels.html.

10. Bruce Wells, "H.L. Hunt and the East Texas Oilfield," American Oil & Gas Historical Society, accessed December 5, 2018, https://aoghs .org/petroleum-pioneers/east-texas-oilfield/.

11. Ryohei Nakagawan, "Japan-U.S. Trade and Rethinking the Point of No Return toward the Pearl Harbor, " *Ritsumeikan Annual Review of International Studies* 9 (2010): 101–123, http://www.ritsumei.ac.jp/acd /cg/ir/college/bulletin/e-vol.9/06Ryohei%20Nakagawa.pdf.

12. "The Aluminum Industry—Part I: Development of Production," Federal Reserve Bank of San Francisco, Monthly Review, August 1957.

13. "Alcoa Inc. Needed Electricity," City of Alcoa, Tennessee, https:// www.cityofalcoa-tn.gov/188/Alcoa-Inc-Needed-Electricity, accessed May 27, 2018. According to p. 512 of the *Statistical Abstract of the United States*, 1949 edition, the annual electricity consumption in 1943 in the United States was approximately 267 billion kilowatt-hours. The 22 billion kilowatt-hours consumed by aluminum smelters that year was about 8 percent of national consumption. https://www.census.gov/library /publications/1949/compendia/statab/70ed.html, accessed May 27, 2018.

14. "Mystery Town Cradled Bomb," *Life Magazine*, August 20, 1945.

15. US Department of Energy, "The War Effort in East Tennessee," K-25 Virtual Museum, http://www.k-25virtualmuseum.org/site-tour/the -war-effort-in-east-tennessee.html, accessed July 22, 2018.

16. B. Cameron Reed, "Kilowatts to Kilotons: Wartime Electricity Use at Oak Ridge," *History of Physics Newsletter* 12, no. 6 (Spring 2015), https://www.aps.org/units/fhp/newsletters/spring2015/upload/spring15 .pdf, accessed May 27, 2018.

17. "B Reactor," US Department of Energy, accessed December 5, 2018, https://www.energy.gov/management/b-reactor.

18. David Halberstam, *The Fifties* (New York: Random House, 1993).

19. International Energy Agency, *Energy Supply Security: Emergency Response of IEA Countries 2014* (Paris, 2014).

20. Ibid.

21. Ibid.

22. Maxim Tucker, "Coal Cutoff Escalates Russia-Ukraine Tensions," Politico, November 27, 2015.

23. Nolan Peterson, "Ukraine Turns to American Coal to Defend Itself Against Russia," *Daily Signal*, November 9, 2017.

24. Sheryl Gay Stolberg, "Saudis Rebuff Bush's Request for More Oil Production," *New York Times*, May 16, 2008.

25. Associated Press, "Saudi Arabia Rebuffs Bush on Oil Production," May 16, 2018.

26. Summer Said, Mark H. Anderson, and Peter Nicholas, "Trump Asks Saudi Arabia to Pump More Oil, Citing High Prices," *Wall Street Journal*, June 30, 2018.

27. David Ebner and Barrie McKenna, "The Pipes That Bind," *Globe and Mail*, February 29, 2008, https://www.theglobeandmail.com/report-on-business/the-pipes-that-bind/article18138386/, accessed June 1, 2018.

28. Gal Luft, "When Hannibal Met Heidi," *Chicago Tribune*, August 25, 2009.

29. Jason Allardyce, "Lockerbie Bomber 'Set Free for Oil,'" *Sunday Times*, August 30, 2009.

30. Luft, "When Hannibal Met Heidi."

31. Alexandra Phillips, "The US Shale Boom," *Harvard International Review*, June 14, 2014.

32. US Department of Energy, Energy Information Administration, "World Oil Transit Chokepoints," July 25, 2017.

33. Associated Press, "Iran Boats Harassed US Navy Ships in Strait of Hormuz, Pentagon Says," January 6, 2008.

34. Mohammed Ibrahim and Graham Bowley, "Pirates Say They Freed Saudi Tanker for $3 Million," *New York Times*, January 9, 2009.

35. Brandon Prins, Ursula Daxecker, and Anup Phayal, "Somali Pirates Just Hijacked an Oil Tanker. Here's What Pirates Want—and Where They Strike," *Washington Post*, March 14, 2017, https://www.washingtonpost.com/news/monkey-cage/wp/2017/01/25/what-do-pirates-want-to-steal-riches-at-sea-so-they-can-pay-for-wars-on-land/?noredirect=on&utm_term=.402f92c7c2d9, accessed June 10, 2017.

36. Frank Holmes, "The Somali Pirate Attacks Are Taking a Toll on Oil Prices," *Business Insider*, May 21, 2011.

37. Christopher Helman, "For US Military, More Oil Means More Death," *Forbes*, November 12, 2009.

38. Report of the Defense Science Board Task Force on DoD Energy Strategy, "More Fight—Less Fuel," Office of the Under Secretary of

Defense for Acquisition, Technology and Logistics (ATL), Arlington, VA, February 2008.

39. Jerry Warner and P.W. Singer, "Fueling the 'Balance': A Defense Energy Strategy Primer," Brookings Institution, August 25, 2009.

40. Report of the Defense Science Board Task Force, "More Capable Warfighting Through Reduced Fuel Burden," Office of the Under Secretary of Defense for Acquisition, Technology and Logistics (ATL), May 2001.

41. US Department of Defense, Air Force Research Lab, via Lt. Gen (ret.) Ken Eickmann, personal communication.

42. Christopher Helman, "For US Military, More Oil Means More Death," *Forbes*, November 12, 2009.

43. Deloitte, "Casualty Costs of Fuel and Water Resupply Convoys in Afghanistan and Iraq," February 25, 2010, https://www.army-technology .com/features/feature77200/, accessed July 22, 2018.

44. "Exploding Fuel Tankers Driving US Army to Solar Power," *Bloomberg*, October 1, 2013, https://fuelfix.com/blog/2013/10/01 /exploding-fuel-tankers-driving-u-s-army-to-solar-power/.

45. Michael E. Webber, "Conflict Between Russia and Georgia Adds New Twist to the Energy War," *Austin American-Statesman*, August 17, 2008.

46. "NATO Carries Out Its Most Intensive Bombing over Last 24 Hours," CNN, May 14, 1999, http://www.cnn.com/WORLD/europe /9905/14/kosovo.01/, accessed May 20, 2018.

47. "Fact File: Blackout Bombs," BBC News, http://news.bbc.co.uk/2 /hi/americas/2865323.stm, accessed May 20, 2018.

48. Ibid.

49. Report of the Defense Science Board Task Force on DoD Energy Strategy, "More Fight—Less Fuel."

50. Naureen S. Malik, "Can America's Power Grid Survive an Electromagnetic Attack?," *Bloomberg*, December 22, 2007, https://www .bloomberg.com/news/articles/2017-12-22/hardening-power-grids-for -nuclear-and-emp-attacks-by-north-korea, accessed July 22, 2018.

51. Personal communication from Evan Pierce, ERCOT black start training, University of Texas at Austin, May 2, 2018.

52. Michael E. Webber, "Breaking the Energy Barrier," *Earth*, September 2009.

53. US Postal Services, "Postal Facts," https://facts.usps.com/size -and-scope/; and "Top Twelve Things You Should Know About the

US Postal Service," https://facts.usps.com/top-facts/#fact93, accessed June 3, 2018.

54. Report of the Defense Science Board Task Force on DoD Energy Strategy, "More Fight—Less Fuel," Office of the Under Secretary of Defense for Acquisition, Technology & Logistics (ATL), February 2008.

55. John Broder, "Climate Change Seen as Threat to US Security," *New York Times*, August 8, 2009.

56. Michael E. Webber, "Why the Withering Nuclear Power Industry Threatens US National Security," *The Conversation*, August 10, 2017.

57. Report of the Defense Science Board Task Force on DoD Energy Strategy, "More Fight—Less Fuel"; Strategic Environmental Research and Development Program (SERDP) and Environmental Security Technology Certification Program (ESTCP), "2014 QDR Emphasizes Climate Change and Energy," April 14, 2014, https://www.serdp-estcp.org/News -and-Events/Blog/2014-QDR-Emphasizes-Climate-Change-and-Energy, accessed July 23, 2018.

58. Morgan Bazilian, personal communication and public tweet, July 21, 2018, https://twitter.com/MBazilian/status/1020719142645608454, accessed July 22, 2018.

## EPILOGUE: THE FUTURE OF ENERGY

1. This concept of an *energy-lover's dilemma* emerged from a conversation with Russell Gold, senior energy reporter at the *Wall Street Journal*, who used the phrase "fossil fuel–lover's dilemma."

2. Wolfgang Schivelbusch, *Disenchanted Night: The Industrialization of Light in the Nineteenth Century* (Berkeley: University of California Press, 1995), p. 74.

3. Richard E. Smalley, "Future Global Energy Prosperity: The Terawatt Challenge," *MRS Bulletin* 30 (June 2005).

4. Evelyn C. Webber, "Why We Can't Use Corn!," personal essay, February 11, 2008.

5. This anecdote was first published in Michael E. Webber, "Three Cheers for Peak Oil!" *Earth*, June 2009.

6. Michael E. Webber and Sheril R. Kirshenbaum, "It's Time to Shine the Spotlight on Energy Education," *The Chronicle of Higher Education*, January 22, 2012.

7. John Cook, Peter Ellerton, and David Kinkead, "Deconstructing Climate Misinformation to Identify Reasoning Errors," *Environmental Research Letters*, February 6, 2018.

# Index

Michael E. Webber is the Josey Centennial Professor in Energy Resources and professor of mechanical engineering at the University of Texas at Austin. He is also the author of *Thirst for Power*. He lives in Paris, France, where he is serving as the Chief Science and Technology Officer for Engie, a global energy and infrastructure services firm.